Maynard's English Classic Series

LYELL'S
TRAVELS IN NORTH AMERICA

IN THE YEARS 1841-2

ABRIDGED AND EDITED BY JOHN P CUSHING, Ph. D.,
HEAD MASTER OF THE NEW HAVEN HIGH SCHOOL,
SOMETIME PROFESSOR OF ECONOMICS IN KNOX COLLEGE

NEW YORK

CHARLES E. MERRILL CO.

44–60 East Twenty-third Street

CONTENTS

CHAPTER XIV

CHAPTER XV

CHAPTER XVI

CHAPTER XVII

CHAPTER XVIII

INTRODUCTION

Sir Charles Lyell, a noted English geologist, was born at Kinnordy, Forfarshire, Scotland, November 14, 1797, and died in London, February 22, 1875. His boyhood home in the New Forest gave him large opportunities for the cultivation of the natural sciences, toward which he had a strong inclination He was a student in Exeter College, Oxford, and was graduated in 1821. He immediately began the study of law, entering Lincoln's Inn, and in 1825 was called to the bar. But his favorite science drew him away from the legal profession, and he became a geologist, a friend and companion of his former Oxford professor of geology, Dr. Buckland. In 1819 he was elected a member of the Linnean and Geological Societies; and in 1822 he read his first paper, "On the Marls of Forfarshire," before the latter society. In 1823 he went to France with introductions to Cuvier, Humboldt and other men of science; and in 1824 he made a geological tour in Scotland in company with Dr. Buckland. In 1826 he was elected fellow of the Royal Society, from which in later years he received its highest honors, the Copley and the Royal Medals.

His principal work, *The Principles of Geology*, has as a secondary title "An Attempt to Explain the Former Changes of the Earth's Surface by Reference to Causes Now in Operation." This was the theme to which he devoted his life. Between 1830, the year of the appearance of the first volume of the " Principles," and 1872, eleven editions were published; and only a few days before his death, he finished revising the twelfth edition, which appeared in 1876. *The Elements of Geology*, published in 1836, went through six editions in the author's lifetime The *Student's Manual of the Elements of Geology* was based upon this latter work. *The An-*

9

tiquity of Man, published in 1863, ran through three editions in one year.

Lyell held for a short time in 1831 the professorship of geology at King's College, London; but whenever his literary duties would permit, he lost no opportunity of exploring distant lands in the interests of science. In 1834 he made an excursion to Denmark and Sweden, and again in 1837 And besides the British Isles, his investigations carried him to Belgium, Switzerland, Germany, Spain, Madeira, Teneriffe, Sicily and the United States.

Lyell visited this country in 1841, having been invited to deliver the Lowell Lectures in Boston. He remained here until the Fall of the following year, and took the opportunity of traveling widely over a large portion of the northern and middle states. His work, *Travels in North America in the Years 1841–2*, gives one a good account of men and manners in this country from the viewpoint of a foreigner. He again visited this country in 1845, and left his account of the trip in *A Second Visit to the United States*.

Lyell received the honor of Knighthood in 1848; and in the year 1864, when he was president of the British Association, he was created a baronet. In his advanced years, his sight, always feeble, failed him altogether. He died, as has been stated, in 1875; and was buried in Westminister Abbey. He holds first rank among geologists; and belongs to that group of great English scientists, among whom were Darwin, Huxley, Tyndall and Spencer.

In the present edition, the technical geological portions of the work have been omitted; yet it is hoped that this volume will not remind one of the play of *Hamlet* with Hamlet left out. There is so much of genuine interest here, there are so many reflections upon the political, social and industrial life of the new nation, that it is believed that this will be a stimulating work to place in any student's hands. The quaint old-fashioned style will be admired for its simplicity, and the retention of the original spelling and punctuation will meet with approval. Lyell observed as a scientist, and wrote as a scientist; and his cautious manner of expressing himself upon doubtful points appears time and again. He states nothing as definite and decided unless his scientific mind has so determined it. The stimulant to the reader comes with the constant flood of contrasts or comparisons that are suggested by the text. Life in 1841 was in almost every respect different from the life of to-day; and the reader will unconsciously turn his thoughts to present day affairs, and compare them with the statements printed in this book. This suggestion of ideas is of great educational importance.

It is interesting to note that at the time Sir Charles Lyell visited this country, Tyler was President. General Harrison had been inaugurated on March 4th, 1841, and died one month later. The population of the United States in 1840 was 17,069,453. There were twenty-six states in the Union; Florida had not been admitted, nor had West Virginia. The most western States were Michigan, Illinois, Missouri, Arkansas and Louisiana. The country was passing through a period of severe financial depression, which began in 1837. Railways had been projected and begun. In 1830 there were four short roads with an aggregate of twenty-three miles. In 1840 there were 2,818 miles of railway. Ocean vessels were successfully propelled by steam in 1838; and the screw-propeller was invented at this time. Friction matches were invented in 1829, and the McCormick reaper in 1834. The wires of the electric telegraph had been strung along Broadway, New York, between 1841 and 1845. Newspapers were beginning to be prominent features in our political life. A group of great Americans,—Hawthorne, Whittier, Longfellow, Bancroft, Emerson and Holmes,—had begun their work; and, in general, the great upward movement, which had begun in Europe in 1832, had spread to this country.

J. P. C.

LYELL'S TRAVELS

LYELL'S TRAVELS

CHAPTER I

Voyage—Harbour of Halifax—Excursions near Boston—Difference of Plants from European Species—Springfield—New Haven—Scenery of the Hudson—Albany—Mohawk Valley—Prosperity and rapid Progress of the People—Lake Ontario.

JULY 20, 1841.—Sailed from Liverpool for Boston, U. S., in the steamship Acadia, which held her course as straight as an arrow from Cape Clear in Ireland to Halifax in Nova Scotia, making between 220 and 280 miles per day.[1]

After the monotony of a week spent on the open sea, we were amused when we came near the great banks which extend from the southern point of Newfoundland, by the rapid passage of the steamer through alternate belts of stationary fog and clear spaces warmed and lighted up with bright sunshine. Looking at the dense fog from the intermediate sunny regions, we could hardly be persuaded that we were not beholding land, so distinct and well-defined was its outline, and such the varieties of light and shade, that some of our Canadian passengers compared it to the patches of cleared and uncleared country on the north shore of the St. Lawrence. These fogs are caused by the meeting, over the great banks, of the warm waters of the gulf stream flowing from the south, and colder currents, often charged with floating ice, from the north, by which, very opposite states in the relative temperature of the sea and atmosphere are produced in spaces closely contiguous. In places where the sea is warmer than the air, fogs are generated.

When the eye has been accustomed for many days to the deep blue of the central Atlantic, the greener tint of the sea over the banks is refreshing. We were within 150 miles of the southern point of Newfoundland when we crossed these banks, over which the shallowest water is said to be about thirty-five fathoms deep. The bottom consists of fine sand, which must be often ploughed up by icebergs, for several of them were seen aground here by some of our passengers on the 31st of July last. The captain tells us that the worst months for crossing the Atlantic to and from Halifax are February and March, and the most agreeable ones, July, August and September. The nearer we approached the American coast, the more beautiful and brilliant were the sunsets. We sometimes compared the changing hues of the clouds and sky to the blue and red colours in a pigeon's neck.

July 31.—On the eleventh day of our voyage we sailed directly into the harbour of Halifax, which by its low hills of granite and slate, covered with birch and spruce fir, reminded me more of a Norwegian Fiord, such as that of Christiania, than of any other place I had seen. I landed here for six hours, with my wife, during which we had time to drive about the town, and see the museum, where I was shown a large fossil tree filled with sandstone, recently sent from strata[1] containing coal in the interior. I resolved to examine these before returning to England, as they appeared, by the description given us, to afford the finest examples yet known in the world of petrified trees occurring in their natural or erect position.

Letters which we had written on the voyage, being now committed to the post-office at Halifax, were taken up next day by the Caledonia steam-ship for England, and in less than a month from the time of our quitting London, our friends in the remote parts of Great Britain (in Scotland and in Devonshire) were reading an account of the harbour of Halifax, of the Micmac Indians with their Esquimaux features, paddling about in canoes of birch bark, and other novelties seen on the shores of the New World. It required the aid of the recently established railroads at home, as well as the Atlantic steam-packets, to render such rapid correspondence possible.

August 2.—A run of about thirty hours carried us to Boston, which we reached in twelve and a half days after leaving Liverpool. The heat here is intense, the harbour and city beautiful, the air clear and entirely free from smoke, so that the shipping may be seen far off, at the end of many of the streets. The Tremont Hotel merits its reputation as one of the best in the world. Recollecting the contrast of everything French when I first crossed the straits of Dover, I am astonished, after having traversed the wide ocean, at the resemblance of every thing I see and hear to things familiar at home. It has so often happened to me in our own island, without travelling into those parts of Wales, Scotland, or Ireland, where they talk a perfectly distinct language, to encounter provincial dialects which it is difficult to comprehend, that I wonder at finding the people here so very English. If the metropolis of New England be a type of a large part of the United States, the industry of Sam Slick,[1] and other writers, in collecting together so many diverting Americanisms and so much original slang, is truly great, or their inventive powers still greater.

I made excursions to the neighborhood of Boston,[2] through Roxbury, Cambridge, and other places, with a good botanist, to whom I had brought letters of introduction. Although this is not the best season for wild flowers, the entire distinctness of the trees, shrubs, and plants, from those on the other side of the Atlantic, affords a constant charm to the European traveller. We admired the drooping American elm, a picturesque tree; and saw several kinds of sumach, oaks with deeply indented leaves, dwarf birches, and several wild roses. Large commons without heaths reminded me of the singular fact that no species of heath is indigenous on the American continent. We missed also the small "crimson-tipped" daisy on the green lawns, and were told that they have been often cultivated with care, but are found to wither when exposed to the dry air and bright sunshine of this climate. When weeds so common with us cannot be reared here, we cease to wonder at the dissimilarity of the native flora of the New World. Yet whenever the aboriginal forests are cleared, we see orchards, gardens, and arable lands, filled with the same fruit trees, the same grain and vegetables, as in Europe, so bountifully has Nature provided that the plants most useful to man should be capable, like himself, of becoming cosmopolites.

Aug. 9.—After a week spent very agreeably at Boston, we started for New Haven in Connecticut, going the first hundred miles on an excellent railway in about five hours, for three dollars each.[1] The speed of the railways in this state, the most populous in the Union, is greater than elsewhere, and I am told that they are made with American capital, and for the most part pay good interest.[2] There are no tunnels, and so few embankments that they afford the traveller a good view of the country. The number of small lakes and ponds, such as are seen in the country between Lund and Stockholm in Sweden, affords a pleasing variety to the scenery, and they are as useful as they are ornamental. The water is beautifully clear, and when frozen to the depth of many feet in winter, supplies those large cubical masses of ice, which are sawed and transported to the principal cities throughout the Union, and even shipped to Calcutta, crossing the equator twice in their outward voyage. It has been truly said, that this part of New England owes its wealth to its industry, the soil being sterile, the timber small, and there being no staple commodities of native growth, except ice and granite.

In the inland country between Boston and Springfield, we saw some sand-hills like the dunes of blown sand near the coast, which were probably formed on the sea-side before the country was elevated to its present height. We passed many fields of maize, or Indian corn, before arriving at Springfield, which is a beautiful village, with fine avenues [1] of the American elm on each side of the wide streets. From Springfield we descended the river Connecticut in a steamboat. Its banks were covered with an elegant species of golden rod, with its showy bright yellow flowers. I have been hitherto disappointed in seeing no large timber, and I am told that it was cut down originally in New England without mercy, because it served as an ambush for the Indians, since which time it has never recovered, being consumed largely for fuel. The Americans of these Eastern States who visit Europe have, strange to say, derived their ideas of noble trees more from those of our principal English parks, than from the native forests of the New World.

The city of New Haven, with a population of 14,000 souls possesses, like Springfield, fine avenues of trees in its streets, which mingle agreeably with the buildings of the university, and the numerous churches, of which we counted near twenty steeples. When attending service, according to the Presbyterian form, in the College chapel on Sunday, I could scarcely believe I was not in Scotland.

The East and West Rocks near New Haven, crowned with trap, bear a strong resemblance to Salisbury

Crags, and other hills of the same structure near Edinburgh. We saw in Hampden parish, lat. 41° 19′, on the summit of a high hill of sandstone a huge erratic block of greenstone, 100 feet in circumference, and projecting 11 feet above the ground. Other large transported fragments have been met with more than 1000 feet above the level of the sea, and everywhere straight parallel furrows appear on the smooth surface of the rocks, where the superficial gravel and sand are removed.

In a garden at New Haven (August 13) I saw, for the first time, a humming bird on the wing. It was fluttering round the flowers of a Gladiolus. In the suburbs we gathered a splendid wild flower, the scarlet Lobelia, and a large sweet-scented water-lily. The only singing bird which we heard was a thrush with a red breast, which they call here the robin. The grasshoppers were as numerous and as noisy as in Italy. As we returned in the evening over some low marshy ground, we saw several fire-flies, showing an occasional bright spark.

Aug. 13.—A large steamer carried us from New Haven to New York, a distance of about ninety miles, in less than six hours. We had Long Island on the one side, and the main land on the other, the scenery at first tame from the width of the channel, but very lively and striking when this became more contracted, and at length we seemed to sail into the very suburbs of the great city itself, passing between green islands, some of them covered with buildings and villas. We had the same bright sunshine which we have enjoyed ever since we landed, and an atmosphere unsullied by the chimnies of countless steam-boats, factories, and houses of a population of more than 300,000 souls, thanks to the remoteness of all fuel save anthracite and wood.

Aug. 16.—Sailed in the splendid new steam-ship the Troy, in company with about 500 passengers, from New York to Albany, 145 miles, at the rate of about 16 miles an hour.[1]   When I was informed that "seventeen of these vessels went to a mile", it seemed incredible, but I found that in fact the deck measured 300 feet in length.   To give a sufficent supply of oxygen to the anthracite, the machinery is made to work two bellows, which blow a strong current of air into the furnace. The Hudson is an arm of the sea or estuary about twelve fathoms deep, above New York, and its waters are inhabited by a curious mixture of marine and fresh-water plants and mollusca.   At first on our left, or on the western bank, we had a lofty precipice of columnar basalt [2] from 400 to 600 feet in height, called the Palisades, extremely picturesque.   This basalt rests on sandstone, which is of the same age as that mentioned before near New Haven, but has an opposite or west-ward dip.   On arriving at the Highlands, the winding channel is closed in by steep hills of gneiss on both sides, and the vessel often holds her course as if bearing directly on the land.   The stranger cannot guess in which direction he is to penetrate the rocky gorge, but he soon emerges again into a broad valley, the blue Catskill mountains appearing in the distance.   The scenery deserves all the praise which has been lavished upon it, and when the passage is made in nine hours it is full of variety and contrast.

At Albany, a town finely situated on the Hudson, and the capital of the State of New York, I found several geologists employed in the Government survey, and busily engaged in forming a fine museum, to illustrate the organic remains and mineral products of the country. This state is divided into about the same number of counties as England, and is not very inferior to it in extent of territory. The legislature four years ago voted a considerable sum of money, more than 200,000 dollars, or 40,000 guineas, for exploring its Natural History and Mineral Structure; and at the end of the first two years several of the geological surveyors, of whom four principal ones were appointed, reported, among other results, their opinion, that no coal would ever be discovered in their respective districts. This announcement caused no small disappointment, especially as the neighboring state of Pennsylvania was very rich in coal. Accordingly, during my tour, I heard frequent complaints that, not satisfied with their inability to find coal themselves, the surveyors had decided that no one else would ever be able to detect any, having had the presumption to pass a sentence of future sterility on the whole land. Yet in spite of these expressions of ill-humour, it was satisfactory to observe that the rashness of private speculators had received a wholesome check; and large sums of money, which for twenty years previously had been annually squandered in trials for coal in rocks below the car-

boniferous series, were henceforth saved to the public.
There can be little doubt that the advantage derived to
the resources of the State by the cessation of this annual
outlay alone, and the more profitable direction since
given to private enterprise, is sufficient to indemnify
the country, on mere utilitarian grounds, for the sum
so munificently expended by the government on geological investigations.

A few years ago it was a fatiguing tour of many weeks
to reach the Falls of Niagara from Albany. We are now
carried along at the rate of sixteen miles an hour, on a
railway often supported on piles, through large swamps
covered with aquatic trees and shrubs, or through dense
forests, with occasional clearings where orchards are
planted by anticipation among the stumps, before they
have even had time to run up a log house. The traveller views with surprise, in the midst of so much unoccupied land, one flourishing town after another, such
as Utica, Syracuse, and Auburn. At Rochester he
admires the streets of large houses, inhabited by 20,000
souls, where the first settler built his log-cabin in the
wilderness only twenty-five years ago. At one point
our train stopped at a handsome new built station-
house, and looking out at one window, we saw a group
of Indians of the Oneida tribe, lately owners of the
broad lands around, but now humbly offering for sale
a few trinkets, such as baskets ornamented with porcu-
pine quills, moccasins of moose-deer skin, and boxes of
birch bark. At the other window stood a well-dressed
waiter handing ices and confectionery. When we re-

flect that some single towns of which the foundations were laid by persons still living, can already number a population, equal to all the aboriginal hunter tribes who possessed the forests for hundreds of miles around, we soon cease to repine at the extraordinary revolution, however much we may commiserate the unhappy fate of the disinherited race. They who are accustomed to connect the romance of their travels in Europe or Asia with historical recollections and the monuments of former glory, with the great masterpieces in the fine arts, or with grand and magnificent scenery, will hardly believe the romantic sensations which may be inspired by the aspect of this region, where very few points of picturesque beauty meet the eye, and where the aboriginal forest has lost its charm of savage wildness by the intrusion of railways and canals. The foreign naturalist indeed sees novelty in every plant, bird, and insect; and the remarkable resemblances of the rocks at so great a distance from home are to him a source of wonder and instruction. But there are other objects of intense interest, to enliven or excite the imagination of every traveller. Here, instead of dwelling on the past, and on the signs of pomp and grandeur which have vanished, the mind is filled with images of coming power and splendour. The vast stride made by one generation in a brief moment of time, naturally disposes us to magnify and exaggerate the rapid rate of future improvement. The contemplation of so much prosperity, such entire absence of want and poverty, so many school-houses and churches, rising everywhere in the woods, and such a general desire of education, with the consciousness that a great continent lies beyond, which has still to be appropriated, fills the traveller with cheering thoughts and sanguine hopes. He may be reminded that there is another side to the picture, that where the success has been so brilliant and where large fortunes have been hastily realised, there will be rash speculations and bitter disappointments; but these ideas do not force themselves into the reveries of the passing stranger. He sees around him the solid fruits of victory, and forgets that many a soldier in the foremost ranks has fallen in the breach; and cold indeed would be his temperament if he did not sympathise with the freshness and hopefulness of a new country, and feel as men past the prime of life are accustomed to feel when in company with the young, who are full of health and buoyant spirits, of faith and confidence in the future.

CHAPTER II

Distant and near View of the Falls of Niagara—Whether the Falls have receded from Queenston to their present Site—Reflections on the Lapse of past Time.

Aug. 27.—We first came in sight of the Falls of Niagara when they were about three miles distant. The sun was shining full upon them—no building in view— nothing but the green wood, the falling water, and the white foam. At that moment they appeared to me more beautiful than I had expected, and less grand; but after several days, when I had enjoyed a nearer view of the two cataracts, had listened to their thundering sound, and gazed on them for hours from above and below, and had watched the river foaming over the rapids, then plunging headlong into the dark pool,— and when I had explored the delightful island which divides the falls, where the solitude of the ancient forest is still unbroken, I at last learned by degrees to comprehend the wonders of the scene, and to feel its full magnificence.

Early in the morning after our arrival, I saw from the window of our hotel, on the American side, a long train of white vapory clouds hanging over the deep chasm below the falls. They were slightly tinted by the rays

of the rising sun, and blown slowly northward by a
gentle breeze from the pool below the cataract, which
was itself invisible from this point of view. No fog was
rising from the ground, the sky was clear above; and as
the day advanced, and the air grew warm, the vapours
all disappeared. The scene reminded me of my first
view of Mount Etna from Catania, at sunrise in the
autumn of 1828, when I saw dense volumes of steam
issuing from the summit of the highest crater in a clear
blue sky, which, at the height of more than two miles
above the sea, assumed at once the usual shape and
hues of clouds in the upper atmosphere. These, too,
vanished before noon, as soon as the sun's heat increased.

Etna presents us not merely with an image of the
power of subterranean heat, but a record also of the
vast period of time during which that power has been
exerted. A majestic mountain has been produced by
volcanic action, yet the time of which the volcano
forms the register, however vast, is found by the geologist to be of inconsiderable amount, even in the modern
annals of the earth's history. In like manner, the Falls
of Niagara teach us not merely to appreciate the power
of moving water, but furnish us at the same time with
data for estimating the enormous lapse of ages during
which that force has operated. A deep and long ravine
has been excavated, and the river has required ages to
accomplish the task, yet the same region affords evidence that the sum of these ages is as nothing, and as
the work of yesterday, when compared to the antece-

dent periods, of which there are monuments in the same district.

It has long been the popular belief, from a mere cursory inspection of this district, that the Niagara once flowed in a shallow valley across the whole platform from the present site of the Falls to the Queenstown heights, where it is supposed the cataract was first situated, and that the river has been slowly eating its way backwards through the rocks for a distance of seven miles. According to this hypothesis, the Falls must have had originally nearly twice their present height, and must have been always diminishing in grandeur from age to age, as they will continue to do in future so long as the retrograde movement is prolonged. It becomes, therefore, a matter of no small curiosity and interest to inquire at what rate the work of excavation is now going on, and thus to obtain a measure for calculating how many thousands of years or centuries have been required to hollow out the chasm already excavated.

It is an ascertained fact, that the Falls do not remain absolutely stationary at the same point of space, and that they have shifted their position slightly during the last half century. Every observer will also be convinced that the small portion of the great ravine, which has been eroded within the memory of man, is so precisely identical in character with the whole gorge for seven miles below, that the river supplies an adequate cause for executing the task assigned to it, provided we grant sufficient time for its completion.

The waters, after cutting through strata of limestone, about fifty feet thick in the rapids, descend perpendicularly at the Falls over another mass of limestone about ninety feet thick, beneath which lie soft shales of equal thickness, continually undermined by the action of the spray driven violently by gusts of wind against the base of the precipice. In consequence of this disintegration, portions of the incumbent rock are left unsupported, and tumble down from time to time, so that the cataract is made to recede southwards. The sudden descent of huge rocky fragments of the undermined limestone at the Horsehoe Falls in 1828, and another at the American Fall in 1818, are said to have shaken the adjacent country like an earthquake. According to the statement of our guide in 1841, Samuel Hooker, an indentation of about forty feet has been produced in the middle of the ledge of limestone at the lesser fall since the year 1815, so that it has begun to assume the shape of a crescent, while within the same period the Horseshoe Fall has been altered so as less to deserve its name. Goat Island has lost several acres in the last four years, and I have no doubt that this waste neither is, nor has been, a mere temporary accident, since I found that the same recession was in progress in various other waterfalls which I visited with Mr. Hall, in the State of New York. Some of these intersect the same rocks as the Niagara—for example, the Genesee at Rochester; others are cutting their way through newer formations as Allan's Creek below Le Roy, or the Genesee at its upper falls at Portage. Mr.

Bakewell calculated that, in the forty years preceding 1830, the Niagara had been going back at the rate of about a yard annually, but I conceive that one foot per year would be a much more probable conjecture, in which case 35,000 years would have been required for the retreat of the Falls from the escarpment of Queenston to their present site, if we could assume that the retrograde movement had been uniform throughout. This, however, could not have been the case, as at every step in the process of excavation the height of the precipice, the hardness of the materials at its base, and the quantity of fallen matter to be removed, must have varied. At some points it must have receded much faster than at present, at others much slower, and it would be scarcely possible to decide whether its average progress has been more or less rapid than now.

Many have been the successive revolutions in organic life, and many the vicissitudes in the physical geography of the globe, and often has the sea been converted into land, and land into sea, since that rock was formed. The Alps, the Pyrenees, the Himalaya, have not only begun to exist as lofty mountain chains, but the solid materials of which they are composed have been slowly elaborated beneath the sea within the stupendous interval of ages here alluded to.

The geologist may muse and speculate on these events until, filled with awe and admiration, he forgets the presence of the mighty cataract itself, and no longer sees the rapid motion of its waters, nor hears their sound, as they fall into the deep abyss. But whenever his thoughts are recalled to the present, the tone of his mind,—the sensations awakened in his soul, will be found to be in perfect harmony with the grandeur and beauty of the glorious scene which surrounds him.

CHAPTER III

*Tour from the Niagara to the Northern Frontier of Pennsylvania
—Scenery—Sudden Growth of New Towns—Humming Birds—
Nomenclature of Places—Refractory Tenants—Travelling in the
States—Politeness of Women—Domestic Service—Progress of Civ-
ilization—Philadelphia—Fire-engines.*

SEPT. 2, 1841.—From Niagara Falls we travelled to
the large town of Buffalo, on the shores of Lake Erie,
and then passed through Williamsville, Le Roy, and
Geneseo, in the State of New York. When at the village of Geneseo, I learnt that ten years before, the bones
of a Mastodon had been obtained from a bog in the
neighbourhood. At the Falls of Le Roy, and at the
Upper Falls of the River Geneseo at Portage, I had
opportunities of observing how both of these cascades
have been cutting their way backwards through the
Silurian rocks, even within the memory of the present
settlers. They have each hollowed out a deep ravine
with perpendicular sides, bearing the same proportion
in volume to the body of water flowing through them
which the great ravine of the Niagara does to that river.

Mr. Hall took leave of us at Geneseo, after which I
set out on a tour to examine the series of rocks between
the upper Silurian strata of the State of New York and
the Coal of Pennsylvania. With this in view I took the

33

direction of Blossberg, where the most northern coal mines of the United States are worked.

On this occasion we left the main road, and entered, for the first time, an American stage-coach, having been warned not to raise our expectations too high in regard to the ease or speed of our conveyance. Accordingly, we found that after much fatigue we had only accomplished a journey of 46 miles in 12 hours. We had four horses; and when I complained at one of the inns that our coachman seemed to take pleasure in driving rapidly over deep ruts and the roughest ground, it was explained to me that this was the first time in his life he had ever attempted to drive any vehicle, whether two or four-wheeled. The coolness and confidence with which every one here is ready to try his hand at any craft is truly amusing. A few days afterwards I engaged a young man to drive me in a gig from Tioga to Blossberg. On the way, he pointed out, first, his father's property, and then a farm of his own, which he had lately purchased. As he was not yet twenty years of age, I expressed surprise that he had got on so well in the world, when he told me that he had been editor of the "Tioga Democrat" for several years, but had now sold his share of the newspaper.

In the region between Lake Erie and the borders of Pennsylvania, as well as in that immediately south of Lake Ontario, there is an entire want of fine scenery, as might have been anticipated where all the strata are horizontal. The monotony of the endless forest is sometimes relieved by a steep escarpment, a river with

wooded islands, or a lake; but the only striking features of the landscape are the waterfalls, and the deep
chasms hollowed out by them in the course of ages. As
the opposite banks of these ravines are on the same
level, including that of the Niagara itself, we come
abruptly to their edges before we have any suspicion
of their existence, and we must travel out of our way
to enjoy a sight of them.

At length we reached the water-shed, where the
streams flow, on one side, northwards to Lake Ontario,
and on the other, southwards, to the Susquehanna. I
began to wonder how the Indians ever obtained any
correct notions of topography in so continuous a forest,
all the smaller rivers, with their islands, being embow-
ered and choked up with trees. I soon ceased to repine
at the havoc that was going on in the fine timber which
bounded our road on every side.

I found a proprietor on Spalding's Creek preparing
to sink a costly shaft for coal, and I earnestly dissuaded
him from his project, referring him to the New York
survey. Every scientific man who discourages a favour-
ite mining scheme must make up his mind to be as ill
received as the physician who gives an honest opinion
that his patient's disorder is incurable.

Sept. 5.—At Bath I hired a private carriage for Corning. Although there are two railways here with Locomotive Engines, one leading to the south, the other
for conveying the coal of Blossberg to the Erie canal,
I looked in vain for the name of Corning in a newly-
published map and was informed that the town was

only two years old. Already the school-house was
finished, the spire of the Methodist church nearly complete, the Presbyterian one in the course of building,
the site of the Episcopalian decided on. Wishing to
have a carriage, I was taken to a large livery stable,
where there were several vehicles and good horses.
The stumps of trees, some six feet high, are still standing in the gardens and between the houses. Our inn-
keeper remarked that the cost of uprooting them would
be nearly equal to that of erecting a log-house on the
same place. I amused myself by counting the rings
of annual growth in these trees, and found that some
had been only forty years old when cut down, yet
when these began to grow, no white man had approached within many leagues of this valley; most of
the older stumps went back no farther than two centuries, or to the landing of the pilgrim fathers, some
few to the time of Sir Walter Raleigh, and scarcely one
to the days of Columbus. I had before remarked that
very ancient trees seemed uncommon in the aboriginal
forests of this part of America. They are usually tall
and straight, with no grass growing under their dark
shade, although the green herbage soon springs up
when the wood is removed and the sun's rays allowed
to penetrate. Some of the stumps, especially those of
the fir tribe, take fifty years to rot away, though exposed in the air to alternations of rain and sunshine, a
fact on which every geologist will do well to reflect,
for it is clear that the trees of a forest submerged beneath the waters, or still more, if entirely excluded

from air, by becoming imbedded in sediment, may en-
dure for centuries without decay, so that there may
have been ample time for the slow petrifaction of erect
fossil trees in the Carboniferous [1] and other formations,
or for the slow accumulation around them of a great
succession of strata.

I asked the landlord of the inn at Corning, who was
very attentive to his guests, to find my coachman. He
immediately called out in the bar-room, "Where is
the gentleman that brought this man here?" A few
days before, a farmer in New York had styled my wife
"the woman," though he called his own daughters
ladies, and would, I believe, have freely extended that
title to their maid-servant. I was told of a witness in
a late trial at Boston, who stated in evidence that
"while he and another gentleman were shovelling up
mud" &c.; from which it appears that the spirit of
social equality has left no other signification to the
terms "gentleman" and "lady" but that of "male and
female individual."

Dr. Saynisch, who was the first to explore the coal
in this region, told me that, soon after he settled here,
he shot a wolf out of his bedroom window. These
animals still commit havoc on the flocks, and last autumn a large panther was killed in the outskirts of
Blossberg, but the bears have not been seen for several
years. We rode in a hot sunny day to a large clearing
in the forest far away from any habitation, and I was
struck with the perfect silence of the surrounding woods.
We heard no call or note of any bird, nothing to remind

us of the chirping of the chaffinch or autumnal song of the robin, the grasshoppers and crickets alone keeping up a ceaseless din day and night. The birds here are very abundant, and some are adorned with brilliant plumage, as the large woodpecker, with its crimson head, the yellow-bird, of the size of a yellow-hammer, with black wings and a bright yellow body, and the red-bird.

A hen humming-bird, far less brilliant in its plumage than the male, flew within a few inches of my face. Its flight and diminutive size reminded me of our humming sphinx, or hawk-moth, like which it remains poised in the air while sucking the flowers, the body seeming motionless, and the wings being invisible from the swiftness of their vibrations. I had before seen one in the wood at Cedarville, sucking the flower of a wild balsam. Dr. Saynisch tells me that on his first visit to these woods, he has known two of these birds at a time to perch on the edge of a cup of water which he held in his hand, and drink without fear. I was aware from Mr. Darwin's Voyage in the Beagle, that in islands like the Galapagos,

"Where human foot hath ne'er or rarely been,"

the wild birds have no apprehension of danger from man; but here, where for ages the Indian hunters preceded the whites, I am surprised to learn that an instinctive dread of the great "usurper" had not become hereditary in the feathered tribe. I was told, however, that in the hunting grounds called Indian

Reservations, within the limits of the settled and civilized states, of which we passed one in New York, the wild animals are comparatively tame, it being a system of the Indians never to molest the game or their prey, except when required for food.

We returned from *Blossberg* by the town of Jefferson, and, sailing down Seneca Lake in a steamboat to Geneva, joined the railway, which carried us back again to Albany. At one of the stations where the train stopped we overheard some young women from Ohio exclaim, "Well, we are in a pretty fix!" and found their dilemma to be characteristic of the financial crisis of these times, for none of their dollar notes of the Ohio banks would pass here.[1] The substantive "fix" is an acknowledged vulgarism, but the verb is used in New England by well-educated people, in the sense of the French "*arranger*" or the English "do." To fix the hair, the table, the fire, means to dress the hair, lay the table, and make up the fire; and this application is, I presume, of Hibernian origin, as an Irish gentleman, King Corney, in Miss Edgeworth's tale of Ormond, says "I'll fix him of his wounds."[2]

There are scarcely any American idioms or words which are not of British origin, some obsolete, others provincial. When the lexicographer, Noah Webster, whom I saw at New Haven, was asked how many new words he had coined, he replied one only, "to demoralize," and that not for his dictionary, but long before in a pamphlet published in the last century.

The nomenclature of the places passed through in

our short excursion of one month was strange enough.
We had been at Syracuse, Utica, Rome, and Parma,
had gone from Buffalo to Batavia and on the same day
breakfasted at St. Helena, and dined at Elba. We
collected fossils at Moscow, and travelled by Painted
Post and Big Flats to Havanna. After returning by
Auburn to Albany, I was taken to Troy, a city of 20,000
inhabitants, that I might see a curious landslip which
had just happened on Mount Olympus, the western
side of that hill, together with a contiguous portion of
Mount Ida, having slid down into the Hudson, and
caused the death of several persons. Fortunately, some
few of the Indian names, such as Mohawk, Ontario,
Oneida, Canandaigua, and Niagara, are retained. Although legislative interference in behalf of good taste
would not be justifiable, Congress might interpose for
the sake of the post-office, and prevent the future mul-
tiplication of the same name for villages, cities, counties,
and townships. An Englishman, it is true, cannot complain, for we follow the same system in our colonies;
and it is high time that the postmaster-general brought
in a bill for prohibiting new streets in London from re-
ceiving names already appropriated and repeated fifty
times in that same city, to the infinite confusion of the
inhabitants and their letter-carriers.

We then made a tour to the Helderberg Mountains,
S. W. of Albany. I rejoiced to see the sugar-maple, an
ornamental tree, spared in the clearings. The sap from
which sugar is made was everywhere trickling down the
wooden troughs from gashes made in the bark.[1] The

red maples were now beginning to assume their bright
autumnal tints, but the rest of the forest was as verdant as ever; a blue Lobelia, which we had gathered
at the Falls of Niagara was still in bloom, together
with many white and blue asters which had only just
come out. The most elegant flower in the woods at
this season is the fringed gentian.

> "Bright with Autumn's dew,
> And colour'd with the Heaven's own blue."

One day at Schoharie, a hawk pounced down from
a lofty tree, and seized a striped squirrel on the ground,
within three yards of our party. It was bearing off
its burden with ease, until, alarmed by our shouts, it
dropped the squirrel, which ran off apparently unhurt.
I observed early in the morning myriads of cobwebs extending from one blade of grass to another, as we often
see them on an English lawn before the dew is dried up.

On our way back from Schoharie to Albany, we found
the country people in a ferment, a sheriff's officer having been seriously wounded when in the act of distrain-
ing for rent, this being the third year of the "Helderberg war," or a successful resistance by an armed
tenantry to the legal demands of their landlord, Mr.
Van Renssalaer. It appears that a large amount of territory on both sides of the river Hudson, now supporting, according to some estimates, a population of 100,000
souls, had long been held in fee by the Van Renssalaer
family, the tenants paying a small ground rent. This
system of·things is regarded by many as not only in-

jurious, because it imposes grievous restraints upon alienation, but as unconstitutional, or contrary to the genius of their political institutions, and tending to create a sort of feudal perpetuity. Some of the leases have already been turned into fees, but many of the tenants were unable or unwilling to pay the prices asked for such conveyances, and declared that they had paid rent long enough, and that it was high time that they should be owners of the land.

A few years ago, when the estates descended from the late General Van Renssalaer to his sons, the attempt to enforce the landlord's rights met with open opposition. The courts of law gave judgment, and the sheriff of Albany having failed to execute his process, at length took military force in 1839, but with no better success. The governor of New York was then compelled to back him with the military array of the state, about 700 men, who began the campaign at a day's notice in a severe snow storm. The tenants are said to have mustered against them 1500 strong, and the rents were still unpaid, when in the following year, 1840, the governor, courting popularity, as it should seem, while condemning the recusants in his message, virtually encouraged them by recommending their case to the favourable consideration of the state, hinting at the same time at legislative remedies. The legislature, however, to their credit, refused to enact these, leaving the case to the ordinary courts of law.

The whole affair is curious, as demonstrating the impossibility of creating at present in this country a class of landed proprietors deriving their income from the letting of lands upon lease. Every man must occupy his own acres. He who has capital enough to stock a farm can obtain land of his own so cheap as naturally to prefer being his own landlord.

Sept. 27, 1841.—We embarked once more on the Hudson, to sail from Albany to New York, with several hundred passengers on board, and thought the scenery more beautiful than ever. The steam-boat is a great floating hotel, of which the captain is landlord. He presides at meals, taking care that no gentlemen take their places at table till all the ladies, or, as we should say in England, the women of every class, are first seated. The men, by whom they are accompanied, are then invited to join them, after which, at the sound of a bell, the bachelors and married men travelling *en garçon* pour into the saloon, in much the same style as members of the House of Commons rush into the Upper House to hear a speech from the throne.

One of the first peculiarities that must strike a foreigner in the United States is the deference paid universally to the sex, without regard to station. Women may travel alone here in stage-coaches, steam-boats, and railways, with less risk of encountering disagreeable behaviour, and of hearing coarse and unpleasant conversation, than in any country I have ever visited. The contrast in this respect between the Americans and the French is quite remarkable   There is a spirit of true gallantry in all this, but the publicity of the railway car, where all are in one long room, and of the large ordinaries, whether on land or water, is a great protection, the want of which has been felt by many a female traveller without escort in England. As the Americans address no conversation to strangers, we soon became tolerably reconciled to living so much in public. Our fellow-passengers consisted for the most part of shopkeepers, artizans, and mechanics, with their families, all well-dressed, and so far as we had intercourse with them, polite and desirous to please. A large part of them were on pleasure excursions, in which they delight to spend their cash.

On one or two occasions during our late tour in the newly-settled districts of New York, it was intimated to us that we were expected to sit down to dinner with our driver, usually the son or brother of the farmer who owned our vehicle. We were invariably struck with the propriety of their manners, in which there was self-respect without forwardness. The only disagreeable adventure in the way of coming into close contact with low and coarse companions, arose from my taking places in a cheap canal-boat near Lockport, partly filled with emigrants, and corresponding somewhat in the rank of its passengers with a third-class railway-carriage in England. I afterwards learnt that I might have hired a good private carriage at the very place where I embarked. This convenience indeed, although there is no posting, I invariably found at my command in all states of the Union, both northern and southern, which I visited during my stay in America.

Travellers must make up their minds, in this as in

other countries, to fall in now and then with free and
easy people. I am bound, however, to say that in the
two most glaring instances of vulgar familiarity which
we have experienced here, we found out that both the
offenders had crossed the Atlantic only ten years before, and had risen rapidly from a humble station.
Whatever good breeding exists here in the middle
classes is certainly not of foreign importation; and
John Bull, in particular, when out of humour with the
manners of the Americans, is often unconsciously beholding his own image in the mirror, or comparing one
class of society in the United States with another in his
own country, which ought, from superior affluence and
leisure, to exhibit a higher standard of refinement and
intelligence.

We have now seen the two largest cities, many towns
and villages, besides some of the back settlements, of
New York and the New England States; an exemplification, I am told, of a population amounting to about
five millions of souls. We have met with no beggars,
witnessed no signs of want, but everywhere the most
unequivocal proofs of prosperity and rapid progress in
agriculture, commerce, and great public works. As
these states are, some of them, entirely free from debt,
and the rest have punctually paid the interest of Government loans, it would be most unjust to apply to
them the disparaging comment "that it is easy to go
ahead with borrowed money." In spite of the constant
influx of uneducated and pennyless adventurers from
Europe, I believe it would be impossible to find five

millions in any other region of the globe whose average moral, social, and intellectual condition stands so high. One convincing evidence of their well-being has not, I think, been sufficiently dwelt upon by foreigners: I allude to the difficulty of obtaining and retaining young American men and women for a series of years in domestic service, an occupation by no means considered as degrading here, for they are highly paid, and treated almost as equals. But so long as they enjoy such facilities of bettering their condition, and can marry early, they will naturally renounce this bondage as soon as possible. That the few, or the opulent class, especially those resident in country places, should be put to great inconvenience by this circumstance, is unavoidable, and we must therefore be on our guard, when endeavouring to estimate the happiness of the many, not to sympathise too much with this minority.

I am also aware that the blessing alluded to, and many others which they enjoy, belong to a progressive, as contrasted with a stationary, state of society;—that they characterize the new colony, where there is abundance of unoccupied land, and a ready outlet to a redundant labouring class. They are not the results of a democratic, as compared with a monarchical or aristocratic constitution, nor the fruits of an absolute equality of religious sects, still less of universal suffrage. Nevertheless, we must not forget how easily all the geographical advantages arising from climate, soil, fine navigable rivers, splendid harbours, and a wilderness in the far West, might have been marred by other laws, and other political institutions. Had Spain colonized this region, how different would have been her career of civilization! Had the puritan fathers landed on the banks of the Plata, how many hundreds of large steamers would ere this have been plying the Parana and Uruguay,—how many railway-trains flying over the Pampas,—how many large schools and universities flourishing in Paraguay!

Sept. 28.—We next went by railway from New York to Philadelphia through the state of New Jersey. Large fields of maize, without the stumps of trees arising above the corn, and villas with neat flower-gardens, seemed a novelty to us after the eye had dwelt for so many hundreds of miles on native forests and new clearings. The streets of Philadelphia rival the finest Dutch towns in cleanliness, and the beautiful avenues of various kinds of trees afford a most welcome shade in summer. We were five days here, and every night there was an alarm of fire, usually a false one; but the noise of the firemen was tremendous. At the head of the procession came a runner blowing a horn with a deep unearthly sound, next a long team of men (for no horses are employed) drawing a strong rope to which the ponderous engine was attached with a large bell at the top, ringing all the way; next followed a mob, some with torches, others shouting loudly; and before they were half out of hearing, another engine follows with a like escort; the whole affair resembling a scene in *Der Freischutz* or *Robert le Diable*, rather than an act in real life. It is, however, no sham, for these young men are ready to risk their lives in extinguishing a fire; and as an apology for their disturbing the peace of the city when there was no cause, we were told "that the youth here require excitement!" They manage these matters as effectively at Boston without turmoil.

CHAPTER V

Wooded Ridges of the Alleghany Mountains—German Patois in Pennsylvania—Election of a Governor at Trenton and at Philadelphia—Journey to Boston—Boston the Seat of Commerce, of Government, and of a University—Lectures at the Lowell Institute—Influence of Oral Instruction in Literature and Science—Blind Asylum—Lowell Factory—National Schools—Society in Boston.

OCTOBER 7, 1841.—The steep slopes, as well as the summits of the ridges in the anthracite region of Pennsylvania, are so densely covered with wood, that the surveyors were obliged to climb to the tops of trees, in order to obtain general views of the country, and construct a geographical map on the scale of two inches to a mile, on which they laid down the result of their geological observations. Under the trees, the ground is covered with Rhododendron, Kalmia, and another evergreen called Sweet Fern, the leaves of which have a very agreeable odour, resembling that of our bog-myrtle, but fainter. The leaves are so like those of a fern or Pteris in form, that the miners call the impressions of the fossil Pecopteris, in the coal-shales "sweet fern."

We found the German language chiefly spoken in this mountainous region, and preached in most of the churches, as at Reading. It is fast degenerating into a patois, and it is amusing to see many Germanized Eng-

lish words introduced even into the newspapers, such
as *turnpeik* for turnpike, *fense* for fence, *flauer* for flour,
or others, such as jail, which have been adopted without alteration.

From the Lehigh Summit Mine, we descended for
nine miles on a railway impelled by our own weight,
in a small car at the rate of twenty miles an hour. A
man sat in front checking our speed by a drag on
the steeper declivities, and oiling the wheels without
stopping. The coal is let down by the same railroad,
sixty mules being employed to draw up the empty cars
every day. In the evening the mules themselves are
sent down standing four abreast and feeding out of
mangers the whole way. We saw them start in a long
train of waggons, and were told, that so completely do
they acquire the notion that it is their business through
life to pull weights up hill, and ride down at their ease,
that if any of them are afterwards taken away from the
mine and set to other occupations, they willingly drag
heavy loads up steep ascents, but obstinately refuse to
pull any vehicle down hill, coming to a dead halt at
the commencement of the slightest slope.

The general effect of the long unbroken summits of
the ridges of the Alleghany Mountains is very monotonous and unpicturesque; but the scenery is beautiful,
where we meet occasionally with a transverse gorge
through which a large river escapes. After visiting the
Beaver Meadow coal field, we left the mountains by
one of these openings, called the Lehigh Gap, wooded
on both sides, and almost filled up by the Lehigh River,

a branch of the Delaware, the banks of which we now followed to Trenton in New Jersey.

On our way, we heard much of a disastrous flood which occurred last spring on the melting of the snow, and swept away several bridges, causing the loss of many lives. I observed the trees on the right bank of the Delaware at an elevation of about twenty-four feet above the present surface of the river, with their bark worn through by the sheets of ice which had been driven against them. The canal was entirely filled up with gravel and large stones to the level of the towing path, twenty feet above the present level of the stream, which appeared to me to be only explicable by supposing the stones to have been frozen and carried by the floating ice.

Oct. 11.—Reaching Trenton, the capital of New Jersey, late in the evening, we found the town in all the bustle of a general election. A new governor and representatives for the State legislature were to be chosen. As parties are nearly balanced, and the suffrage universal, the good order maintained was highly creditable. Processions, called "parades," were perambulating the streets headed by bands of music, and carrying transparencies with lights in them, in which the names of different counties, and mottoes, such as Union, Liberty, and Equality, were conspicuously inscribed. Occasionally a man called out in a stentorian voice, "The ticket, the whole ticket, and nothing but the ticket," which was followed by a loud English hurra, while at intervals a single blow was struck on a great drum, as if to imi-

tate the firing of a gun. On their tickets were printed
the names of the governor, officers, and members for
whom the committee of each party had determined to
vote.

The next day on our return to Philadelphia, we found
that city also in the ferment of an election, bands of
music being placed in open carriages, each drawn by
four horses, and each horse decorated with a flag, attached to its shoulder, which has a gay effect. All day
a great bell tolls at the State House, to remind the
electors of their duties. It sounded like a funeral; and
on my inquiring of a bystander what it meant, one of
the democratic party answered, "It is the knell of the
whigs." In their popular addresses, some candidates
ask the people whether they will vote for the whigs who
will lay on new taxes. As it is well known, that such
taxes must be imposed, if the dividends on the State
bonds are to be paid, these popular appeals are ominous.
The rapid fall in the value of State securities shows that
the public generally have no confidence that the majority of the electors will be proof against the insidious
arts of these demagogues.

Oct. 14.—We came from Philadelphia to Boston,
300 miles, without fatigue in twenty-four hours, by
railway and steam-boat, having spent three hours in
an hotel at New York, and sleeping soundly for six
hours in the cabin of a commodious steamer as we
passed through Long Island Sound. The economy of
time in travelling here is truly admirable. On getting
out of the cars in the morning, we were ushered into a

spacious saloon, where with 200 others we sat down to breakfast, and learnt with surprise, that while thus agreeably employed, we had been carried rapidly in a large ferry-boat without perceiving any motion across a broad estuary to Providence in the State of Rhode Island.

Many trees in New Jersey, Connecticut, and Massachusetts, have now begun to assume their autumnal tints, especially the maples, while the oaks retain their vivid green colour. I can only compare the brightness of the faded leaves, scarlet, purple, and yellow, to that of tulips. It is now the Indian summer, a season of warm sunny weather, which often succeeds to the first frost and rain, a time which the Indians employed in hunting and laying up a store of game for the winter.

Boston, Oct. 14, to Dec. 3, 1841.—It is fortunate that Boston is at once a flourishing commercial port, and the seat of the best endowed university in America, for Cambridge, where Harvard College is situated, is so near, that it may be considered as a suburb of the metropolis. The medical lectures, indeed, are delivered in the city, where the great hospitals are at hand. The mingling of the professors, both literary and scientific, with the eminent lawyers, clergymen, physicians, and principal merchants of the place, forms a society of a superior kind; and to these may be added several persons, who, having inherited ample fortunes, have successfully devoted their lives to original researches in history, and other departments. It is also a political advantage of no small moment that the legis-

lature assembles here, as its members, consisting in
great part of small proprietors farming their own land,
are thus brought into contact with a community in a
very advanced state of civilization, so that they are
under the immediate check of an enlightened public
opinion. It is far more usual to place the capital, as it is
called, in the centre of the State, often in some small
village or town of no importance, and selected from
mere geographical considerations, which might well be
disregarded in a country enjoying such locomotive facilities. An immense sacrifice is then required from
those men of independent fortune, who, for patriotic
motives, must leave the best society of a large city, to
spend the winter in some remote spot in the discharge
of public duties.

I had been invited when in England by Mr. Lowell,
trustee and director of a richly endowed literary and
scientific institution in this city, to deliver a course of
twelve lectures on geology during the present autumn.
According to the conditions of the bequest, the public
have (gratuitous) admission to these lectures; but by
several judicious restrictions, such as requiring ap-
plications for tickets to be made some weeks before,
and compliance with other rules, the trustee has obvi-
ated much of the inconvenience arising from this privilege, for it is well known that a class which pays nothing is irregular and careless in its attendance. As the
number of tickets granted for my lectures amounted
to 4500, and the class usually attending consisted of
more than 3000 persons, it was necessary to divide

them into two sets, and repeat to one of them the next
afternoon the lecture delivered on the previous evening.
It is by no means uncommon for professors who have
not the attraction of novelty, or the advantage which I
happened to enjoy, of coming from a great distance,
to command audiences in this institution as numerous
as that above alluded to. The subjects of their discourses are various, such as natural history, chemistry,
the fine arts, natural theology, and many others.
Among my hearers were persons of both sexes, of every
station in society, from the most (affluent) and eminent
in the various learned professions to the humblest
mechanics, all well dressed and observing the utmost
decorum.

The theatres [1] were never in high favour here, and
most of them have been turned to various secular and
ecclesiastical uses, and among others into lecture rooms,
to which many of the public resort for amusement as
they might formerly have done to a play, after the
labours of the day are over. If the selection of teachers
be in good hands, institutions of this kind cannot fail
to exert a powerful influence in improving the taste and
intellectual condition of the people, especially where
college is quitted at an early age for the business of
active life, and where there is always danger in a commercial community that the desire of money-making
may be carried to excess. It is, moreover, peculiarly
desirable in a democratic state, where the public mind
is apt to be exclusively absorbed in politics, and in a
country where the free competition of rival sects has a

tendency to produce not indifferentism, as some at home may be disposed to think, but too much excitement in religious matters.

We are informed by Mr. Everett, late governor of Massachusetts (since minister of the U. S. in England), that before the existence of the Lowell Foundation, twenty-six courses of lectures were delivered in Boston, without including those which consisted of less than eight lectures, and these courses were attended in the aggregate by about 13,500 persons. But notwithstanding the popularity of this form of instruction, the means of the literary and scientific institutions of the city were wholly inadequate to hold out a liberal and certain reward to men of talent and learning. There were some few instances of continuous courses delivered by men of eminence; but the task more commonly devolved upon individuals who cultivated the art of speaking merely to become the vehicles of second-hand information, and who were not entitled to speak with authority, and from the fulness of their own knowledge.

The rich who have had a liberal education, who know how to select the best books, and can afford to purchase them, who can retreat into the quiet of their libraries from the noise of their children, and, if they please, obtain the aid of private tuition, may doubt the utility of public lectures on the fine arts, history, and the physical sciences. But oral instruction is, in fact, the only means by which the great mass of the middling and lower classes can have their thoughts

turned to these subjects, and it is the fault of the higher
classes if the information they receive be unsound, and
if the business of the teacher be not held in high honour. The whole body of the clergy in every country
and, under popular forms of government, the leading
politicians have been in all ages convinced that they
must avail themselves of this method of teaching, if
they would influence both high and low. No theological dogma is so abstruse, no doctrine of political
economy or legislative science so difficult, as to be
deemed unfit to be preached from the pulpit, or incul-
cated on the hustings. The invention of printing,
followed by the rapid and general dispersion of the
cheap daily newspaper, or the religious tract, have been
by no means permitted to supersede the instrumentality of oral teaching, and the powerful sympathy and
excitement created by congregated numbers. If the
leading patrons and cultivators of literature and physical science neglect this ready and efficacious means of
interesting the multitude in their pursuits, they are
wanting to themselves, and have no right to complain
of the apathy or indifference of the public.

To obtain the services of eminent men engaged in
original researches, for the delivery of systematic
courses of lectures, is impossible without the command
of much larger funds than are usually devoted to this
object. When it is stated that the fees at the Lowell
Institute at Boston are on a scale more than three
times higher than the remuneration awarded to the
best literary and scientific public lecturers in London,

it will at first be thought hopeless to endeavour to carry similar plans into execution in other large cities, whether at home or in the United States. In reality, however, the sum bequeathed by the late Mr. John Lowell for his foundation, though munificent, was by no means enormous, not exceeding much 70,000£., which, according to the usual fate awaiting donations for educational objects, would have been all swallowed up in the erection of costly buildings, after which the learned would be invited to share the scanty leavings of the "Committee of Taste," and the merciless architect, "reliquias Danaum atque inmitis Achillei." [1] But in the present case, the testator provided in his will that not a single dollar should be spent in brick and mortar, in consequence of which proviso, a spacious room was at once hired, and the intentions of the donor carried immediately into effect, without a year's delay.

Mr. John Lowell, a native of Massachusetts, after having carefully studied the educational establishments of his own country, visited London in 1833, and having sojourned there some months, paying a visit to the University of Cambridge and other places, he pursued his travels in the hope of exploring India and China. On his way he passed through Egypt, where, being attacked, while engaged in making a collection of antiques, by an intermittent fever, of which he soon afterwards died, he drew up his last will in 1835, amidst the ruins of Thebes, leaving half of his noble fortune for the foundation of a Literary Institute in his native city.

In the Blind Asylum I saw Laura Bridgman, now in her twelfth year. At the age of two she lost her sight and hearing by a severe illness, but although deaf, dumb, and blind, her mind has been so advanced by the method of instruction pursued by Dr. Howe, that she shows more intelligence and quickness of feeling than many girls of the same age who are in full possession of all their senses. The excellent reports of Dr. Howe, on the gradual development of her mind, have been long before the public, and have recently been cited by Mr. Dickens, together with some judicious observations of his own. Perhaps no one of the cases of a somewhat analogous nature, on which Dugald Stewart and others have philosophised, has furnished so many new and valuable facts illustrating the extent to which all intellectual development is dependent on the instrumentality of the senses in discerning external objects, and, at the same time, in how small a degree the relative acuteness of the organs of sense determine the moral and intellectual superiority of the individual.

Nov. 15.—Went twenty-six miles to the north of Boston, by an excellent railway, to the manufacturing town of Lowell, which has sprung up entirely in the last sixteen years, and now contains about 20,000 inhabitants. The mills are remarkably clean, and well warmed, and almost all for making cotton and woolen goods, which are exported to the West. The young women from the age of eighteen to twenty-five, who attend to the spinning-wheels, are good-looking and neatly dressed, chiefly the daughters of New England farmers, sometimes of the poorer clergy. They belong, therefore, to a very different class from our manufacturing population, and after remaining a few years in the factory, return to their homes, and usually marry. We are told that, to work in these factories is considered far more eligible for a young woman than domestic service, as they can save more, and have stated hours of work (twelve hours a day!), after which they are at liberty. Their moral character stands very high, and a girl is paid off, if the least doubt exists on that point. Boarding-houses, usually kept by widows, are attached to each mill, in which the operatives are required to board; the men and women being separate. This regard for the welfare and conduct of the work-people when they are not on actual duty is comparatively rare in England, where the greater supply of labour would render such interference and kind superintendence much more practicable. Still we could not expect that the results would be equally satisfactory with us, on account of the lower grade of the operatives, and the ignorance of the lower classes in England. In regard to the order, dress, and cleanliness of the people, these merits are also exemplified in the rural districts of Lancashire, and it is usually in our large towns alone, that the work-people are unhealthy and squalid, especially where a number of the poor live crowded together in bad dwellings.

The factories at Lowell are not only on a great scale,

but they have been so managed as to yield high profits, a fact which should be impressed on the mind of every foreigner who visits them, lest, after admiring the gentility of manner and dress of the women and men employed, he should go away with the idea that he had been seeing a model mill, or a set of gentlemen and ladies playing at factory for their amusement. There are few children employed, and those under fifteen are compelled by law to go to school three months in the year, under penalty of a heavy fine. If this regulation is infringed, informers are not wanting, for there is a strong sympathy in the public mind with all acts of the legislature, enforcing education. The Bostonians submit to pay annually for public instruction in their city alone, the sum of 30,000*l.* sterling, which is about equal to the parliamentary grant of this year (1841) for the whole of England, while the sum raised for free schools in this state this year, by taxes for wages of teachers, and their board, and exclusive of funds for building, exceeds 100,000*l.* sterling.

The law ordains, that every district containing fifty families shall maintain one school, for the support of which the inhabitants are required to tax themselves, and to appoint committees annually for managing the funds, and choosing their own schoolmasters. The Bible is allowed to be read in all, and is actually read in nearly all the schools; but the law prohibits the use of books "calculated to favour the tenets of any particular sect of Christians." Parents and guardians are expected to teach their own children, or to procure them to be taught, what they believe to be religious truth, and for this purpose, besides family worship and the pulpit, there are Sunday-schools. The system works well among this church-building and church-going population.

As there is no other region in Anglo-saxondom, containing 750,000 souls, where national education has been carried so far, it is important to enquire to what combination of causes its success is mainly to be attributed. First, there is no class in want or extreme poverty here, partly because the facility of migrating to the west, for those who are without employment, is so great, and also, in part, from the check to improvident marriages, created by the high standard of living to which the lowest work-people aspire, a standard which education is raising higher and higher from day to day. Secondly, I have often heard politicians of opposite parties declare, that there is no safety for the republic, now that the electoral suffrage has been so much extended, unless every exertion is made to raise the moral and intellectual condition of the masses. The fears entertained by the rich of the dangers of ignorance, is the only good result which I could discover tending to counterbalance the enormous preponderance of evil arising in the United States from so near an approach to universal suffrage. Thirdly, the political and social equality of all religious sects,— a blessing which the New Englanders do not owe to the American revolution, for it was fully recognised and enjoyed under the supremacy of the British crown.

This equality tends to remove the greatest stumbling block, still standing in the way of national instruction in Great Britain, where we allow one generation after another of the lower classes to grow up without being taught good morals, good behaviour, and the knowledge of things useful and ornamental, because we cannot all agree as to the precise theological doctrines in which they are to be brought up. Politically, all sects in Massachusetts are ready to unite against the encroachments of any other, and a great degree of religious freedom is enjoyed, in consequence of there being no sect to which it is ungenteel to belong, no consciences sorely tempted by ambition to conform to a more fashionable creed.

The twenty-fifth of November was appointed by the Governor of the State to be what is here called Thanksgiving-Day, an institution as old as the times of the Pilgrim Fathers, one day in the year being set apart for thanksgiving for the mercies of the past year. As a festival it stands very much in the place of Christmas Day as kept in England and Germany, being always in winter, and every body going to church in the morning and meeting in large family parties in the evening. To one of these we were most kindly welcomed; and the reception which we met with here and in the few families to which we had letters of introduction, made us entirely forget that we were foreigners. Several of our new acquaintances indeed had travelled in England and on the Continent, and were in constant correspondence with our own literary and scientific friends, so that we were always hearing from them some personal news of those with whom we were most intimate in Europe, and we often reflected with surprise in how many parts of England we should have felt less at home.

I remember an eminent English writer once saying to me, when he had just read a recently-published book on the United States, "I wonder the author went so far to see disagreeable people, when there are so many of them at home." It would certainly be strange if persons of refined habits, even without being fastidious, who travel to see life, and think it their duty, with a view of studying character, to associate indiscriminately with all kinds of people, visiting the first strangers who ask them to their houses, and choosing their companions without reference to congeniality of taste, pursuits, manners, or opinions, did not find society in their own or any other country in the world intolerable.

CHAPTER VI

Fall of Snow and Sleigh-driving at Boston—Journey to New Haven—Income of Farmers—Baltimore—Washington—Natural Impediments to the Growth of Washington—Why chosen for the Capital—Richmond.

Nov. 29, 1841.—Although we were in the latitude of Rome, and there were no mountains near us, we had a heavy fall of snow at Boston this day, followed by bright sunshine and hard frost. It was a cheerful scene to see the sleighs gliding noiselessly about the streets, and to hear the bells, tied to the horses' heads, warning the passer-by of their swift approach. As it was now the best season to geologise in the southern States, I determined to make a flight in that direction; and we had gone no farther than New Haven before we found that all the snow had disappeared. I accordingly took the opportunity when there of making a geological excursion, with Mr. Silliman, jun., Professor Hubbard, and Mr. Whelpley, to examine the red sandstone strata by the side of a small waterfall, one mile from Durham, in Connecticut.

In the neighborhood of Durham we learnt that a snowstorm, which occurred there in the first week of October, had seriously injured the woods, weighing down the boughs then in full leaf, and snapping off

65

the leading shoots. For the first time in the United States I heard great concern expressed for the damage sustained by the timber, which is beginning to grow scarce in New England, where coal is dear.

The valley of the Connecticut presents a pleasing picture of a rural population, where there is neither poverty nor great wealth. I was told by well-informed persons, that if the land and stock of the farmers or small proprietors were sold off and invested in securities giving six per cent interest, their average incomes would not exceed more than from 80£ to 120£ a year. An old gentleman who lately revisited Durham, his native place, after an absence of twenty-five years, told me that in this interval the large families, the equal subdivision of the paternal estates among children, and the efforts made for the outfit of sons migrating to the West, had sensibly lowered the fortunes of the Connecticut yeomanry, so that they were reduced nearer to the condition of labourers than when he left them.

Pursuing my course southwards, I found that the snowstorm had been less heavy at New York, still less at Philadelphia, and after crossing the Susquehanna (Dec. 13) the weather began to resemble that of an English spring. In the suburbs of Baltimore, the locomotive engines being detached, our cars were drawn by horses on a railway into the middle of the town. Maryland was the first slave state we had visited; and at Baltimore we were reminded for the first time of the poorer inhabitants of a large European city by the mean dwellings and dress of some of the labouring class, both coloured and white.

At Washington I was shown the newly-founded national museum, in which the objects of natural history and other treasures collected during the late voyage of discovery to the Antarctic regions, the South Seas, and California, are deposited. Such a national repository would be invaluable at Philadelphia, New York, or Boston, but here there is no university, no classes of students in science or literature, no philosophical societies, no people who seem to have any leisure. The members of Congress rarely have town residences in this place, but, leaving their families in large cities, where they may enjoy more refined society, they live here in boarding-houses until their political duties and the session are over. If the most eminent legislators and statesmen, the lawyers of the supreme courts, and the foreign ambassadors, had all been assembled here for a great part of the year with their families, in a wealthy and flourishing metropolis, the social and political results of a great centre of influence and authority could not have failed to be most beneficial. Circumstances purely accidental, and not the intentional jealousy of the democracy, have checked the growth of the capital, and deprived it of the constitutional ascendency which it might otherwise have exerted. Congress first assembled in Philadelphia, where the declaration of independence was signed; but after the close of the revolutionary war in June, 1783, a party of the disbanded army marched

to that city to demand their arrears of pay, surrounding the building in which the representatives of the
people were sitting, with fixed bayonets for about
three hours. This alarm caused them to adjourn and
meet at Princeton, New Jersey, and afterwards to seek
some permanent seat of government. But for this
untoward event, Philadelphia might have remained
the federal metropolis, and in that case would certainly have lifted up her head above other cities in the
New World—

"Quantum lenta solent inter viburna cupressi." [1]

General Washington is said to have selected the
present site of the capital as the most central spot
on the Atlantic border, being midway between Maine
and Florida, and being also at the head of the naviga-
tion of a great river. He had observed that all the
other principal cities eastward of the Alleghany mountains had sprung up on similar sites; but unfortunately
the estuary of the Potomac is so long and winding, that
to ascend from its mouth to Washington is said often
to take a vessel as long as to cross from Liverpool to
the mouth of the river. Had Annapolis, which is only
thirty miles distant, been chosen as the capital, it is
believed that it would, ere this, have contained 100,000
inhabitants.

We were present at an animated debate in the House
of Representatives, on the proposed protective tariff,
and a discussion in the senate on "Ways and Means,"
both carried on with great order and decorum. After

being presented to the President, and visiting several persons to whom we had letters, we were warned by a slight sprinkling of snow that it was time to depart and migrate further southwards. Crossing the Potomac, therefore, I proceeded to Richmond, in Virginia, where I resolved to sail down the James River, in order to examine the geology of the tertiary strata on its shores.

On entering the station-house of the railway which was to carry us to our place of embarkation, we found a room with only two chairs in it. One of these was occupied by a respectable-looking woman, who immediately rose, intending to give it up to me, an act betraying that she was English, and newly-arrived, as an American gentleman, even if already seated, would have felt it necessary to rise and offer the chair to any woman, whether mistress or maid, and she, as a matter of course, would have accepted the proffered seat. After I had gone out, she told my wife that she and her husband had come a few months before from Hertfordshire, hoping to get work in Virginia, but she had discovered that there was no room here for poor white people, who were despised by the very negroes if they laboured with their own hands. She had found· herself looked down upon, even for carrying her own child, for they said she ought to hire a black nurse. These poor emigrants were now anxious to settle in some free state.

As another exemplification of the impediments to improvement existing here, I was told that a New England agriculturalist had bought a farm on the south side of the James river, sold off all the slaves, and introduced Irish labourers, being persuaded that their services would prove more economical than slave-labour. The scheme was answering well, till, by the end of the third year, the Irish became very much dissatisfied with their position, feeling degraded by losing the respect of the whites, and being exposed to the contempt of the surrounding negroes. They had, in fact, lowered themselves by the habitual performance of offices which, south of the Potomac, are assigned to hereditary bondsmen.

All the planters in this part of Virginia, to whose houses I went without letters of introduction, received me politely and hospitably. To be an Englishman engaged in scientific pursuits was a sufficient passport, and their servants, horses, and carriages, were most liberally placed at my disposal.

I crossed to the north side of the James river, being rowed out at sunrise far from the shore to wait for a steamer. The hour of her arrival being somewhat uncertain, we remained for some time in the cold, muffled up in our cloaks, in a small boat moored to a single wooden pile driven into a shoal, with three negroes for our companions. The situation was desolate in the extreme, both the banks of the broad estuary appearing low and distant, and as wild and uninhabited as when first discovered in 1607, by Captain Smith, before he was taken prisoner, and his life saved by the Indian maiden Pocahontas. At length we gladly hailed the large steamer as she came down rapidly towards us, and my luggage was immediately taken charge of by two of the sable crew.

We disembarked in a few hours near the old deserted village of Jamestown, at the Grove Landing, seven miles south of Williamsburg. We then visited Williamsburg, where there is a University founded by William and Mary, and therefore very ancient for this country.

CHAPTER VII

Dec. 23, 1841.—From Williamsburg we went to Norfolk in Virginia, passing down the James river in a steamer, and from Norfolk by railway to Weldon in North Carolina, passing for eighty miles through a low level country, covered with fir trees, and called the Pine Barrens. On our way we were overtaken by rain, which turned to sleet, and in the evening formed a coating of ice on the rails, so that the wheels of the engine could not take hold. There was a good stove and plenty of fuel in the car, but no food. After a short pause, the engineer backed the locomotive for half a mile over that part of the rail from which the snow and ice had just been brushed and scraped away by the passage of the train; then, returning rapidly, he gained sufficient momentum to carry us on two or three miles farther, and, by several repetitions of this manoeuvre, he brought us, about nightfall, to a small watering station, where there was no inn, but a two-storied cottage not far off.

Here we were made welcome, and as we had previously dropped by the way all of our passengers except two, were furnished with a small room to ourselves, and a clean comfortable bed. We soon made a blazing wood-fire, and defied the cold, although we could see plainly the white snow on the ground through openings in the unplastered laths of which the walls of the house were made. Before morning all the snow was melted, and we again proceeded on our way through the Pine Barrens.

Our car, according to the usual construction in this country, was in the shape of a long omnibus, with the seats transverse, and a passage down the middle, where, to the great relief of the traveller, he can stand upright with his hat on, and walk about, warming himself when he pleases at the stove, which is in the centre of the car. There is often a private room fitted up for the ladies, into which no gentleman can intrude, and where they are sometimes supplied with rocking-chairs, so essential to the comfort of the Americans, whether at sea or on land, in a fashionable drawing-room or in the cabin of a ship. It is singular enough that this luxury, after being popular for ages all over Lancashire, required transplantation to the New World before it could be improved and become fashionable, so as to be reimported into its native land.

The Pine Barrens, on which the long-leafed or pitch pines flourish, have for the most part a siliceous soil, and form a broad belt many hundred miles in length, running parallel to the coast, in the region called the Atlantic Plain, before alluded to. The sands cause swamps where peculiar kinds of evergreen oaks, the

cypress or cedar, tall canes and other plants abound.
Many climbers called here wild vines, encircle the trunks
of the trees, and on the banks of the Roanoke, near
Weldon, I saw numerous missletoes with their white
berries. The Pine Barrens retain much of their verdure
in winter, and were interesting to me from the uniform-
ity and monotony of their general aspect, for they con-
stitute, from their vast extent, one of the marked features in the geography of the globe, like the Pampas of
South America.

There are many swamps or morasses in this low flat
region, and one of the largest of these occurs between
the towns of Norfolk and Weldon. We traversed several miles of its northern extremity on the railway,
which is supported on piles. It bears the appropriate
and very expressive name of the "Great Dismal," and
is no less than forty miles in length from north to south,
and twenty-five miles in its greatest width from east
to west, the northern half being situated in Virginia,
the southern in North Carolina. I observed that the
water was obviously in motion in several places, and
the morass has somewhat the appearance of a broad
inundated river-plain, covered with all kinds of aquatic
trees and shrubs, the soil being as black as in a peat-
bog.

It is one enormous quagmire, soft and muddy, except where the surface is rendered partially firm by a
covering of vegetables, and their matted roots; yet,
strange to say, instead of being lower than the level of
the surrounding country, it is actually higher than

nearly all the firm and dry land which encompasses it, and, to make the anomaly complete, in spite of its semi-fluid character, it is higher in the interior than towards its margin.

The only exceptions to both these statements is found on the western side, where, for instance, for the distance of about twelve or fifteen miles, the streams flow from slightly elevated but higher land, and supply all its abundant and overflowing water. Towards the north, the east, and the south, the waters flow from the swamp to different rivers, which give abundant evidence, by the rate of descent, that the Great Dismal is higher than the surrounding firm ground. This fact is also confirmed by the measurements made in levelling for the railway from Portsmouth to Suffolk, and for two canals cut through different parts of the morass, for the sake of obtaining the timber. The railway itself, when traversing the Great Dismal, is literally higher than when on the land some miles distant on either side, and is six to seven feet higher than where it passes over dry ground, near to Suffolk and Portsmouth. Upon the whole the centre of the morass seems to lie more than twelve feet above the flat country round it. The soil of the swamp is formed of vegetable matter, usually without any admixture of earthy particles. We have here, in fact, a deposit of peat from ten to fifteen feet in thickness, in a latitude where, owing to the heat of the sun, and the length of the summer, no peat mosses like those of Europe would be looked for under ordinary circumstances.

In countries like Scotland and Ireland, where the climate is damp, and the summer short and cool, the natural vegetation of one year does not rot away during the next in moist situations. If water flows into such land, it is absorbed, and promotes the vigorous growth of mosses and other aquatic plants, and when they die, the same water arrests their putrefaction. But as a general rule, no such accumulation of peat can take place in a country like that of Virginia, where the summer's heat causes annually as great a quantity of dead plants to decay as is equal in amount to the vegetable matter produced in one year.

It has already been stated that there are many trees and shrubs in the region of the Pine Barrens (and the same may be said of the United States generally), which, like our willows, flourish luxuriously in water. The juniper trees, or white cedar, stand firmly in the softest parts of the quagmire, supported by their long tap roots, and afford, with many other evergreens, a dark shade, under which a multitude of ferns, reeds, and shrubs, from nine to eighteen feet high, and a thick carpet of mosses, four or five inches high, spring up and are protected from the rays of the sun. When these are most powerful, the large cedar and many other deciduous trees are in full leaf. The black soil formed beneath this shade, to which the mosses and leaves make annual additions, does not perfectly resemble the peat of Europe, most of the plants being so decayed as to leave little more than soft black mud, without any traces of organization. This loose soil is called sponge by the labourers; and it has been ascertained that, when exposed to the sun, and thrown out on the banks of the canal, where clearings have been made, it rots entirely away. Hence it is evident that it owes its preservation in the swamp to moisture and the shade of the dense foliage. The evaporation continually going on in the wet spongy soil during summer cools the air, and generates a temperature resembling that of a northern climate, or a region more elevated above the level of the sea.

Numerous trunks of large and tall trees lie buried in the black mire of the morass. In so loose a soil they are easily overthrown by winds, and nearly as many have been found lying beneath the surface of the peaty soil, as standing erect upon it. When thrown down, they are covered by water, and keeping wet they never decompose, except the sap wood, which is less than an inch thick. Much of the timber is obtained by sounding a foot or two below the surface, and it is sawn into planks while half under water.

The bears inhabiting the swamp climb trees in search of acorns and gum berries, breaking off large boughs of the oaks in order to draw the acorns near to them. These same bears are said to kill hogs and even cows. There are also wild cats, and occasionally a solitary wolf, in the morass.

CHAPTER VIII

*Tour to Charleston, South Carolina—Facilities of Locomotion—
Augusta—Voyage down the Savannah River—Fever and Ague—
Pine Forests of Georgia—Alligators and Land-Tortoises—
Warmth of Climate in January—Passports required of Slaves.*

Dec. 28.—Charleston, South Carolina. We arrived
here after a journey of 160 miles through the pine
forests of North Carolina, between Weldon and Wilmington, and a voyage of about 17 hours, in a steam
ship, chiefly in the night between Wilmington and this
place. Here we find ourselves in a genial climate, where
the snow is rarely seen, and never lies above an hour
or two upon the ground. The rose, the narcissus, and
other flowers, are still lingering in the gardens, the
woods still verdant with the magnolia, live oak, and
long-leaved pine, while the dwarf fan palm or palmetto,
frequent among the underwood, marks a more southern
region. In less than four weeks since we left Boston, we
have passed from the 43d to the 33d degree of latitude,
carried often by the power of steam for several hundred
miles together through thinly peopled wildernesses, yet
sleeping every night in good inns, and contrasting the
facilities of locomotion in this new country with the
difficulties we had contended with the year before when

78

travelling in Europe, through populous parts of Touraine, Brittany, and other provinces of France.

At Charleston I made acquaintance with several persons zealously engaged in the study of Natural History, and then went by an excellent railway 136 miles through the endless pine woods to Augusta, in Georgia. This journey, which would formerly have taken a week, was accomplished between sunrise and sunset, and, as we scarcely saw by the way any town or village, or even a clearing, nor any human habitation except the station houses, the spirit of enterprise displayed in such public works filled me with astonishment which increased the farther I went South. Starting from the sea-side, and imagining that we had been on a level the whole way, we were surprised to find in the evening, on reaching the village of Aiken, sixteen miles from Augusta, that we were on a height several hundred feet above the sea, and that we had to descend a steep inclined plane to the valley of the Savannah river.

I had been warned by my scientific friends in the North that the hospitality of the planters might greatly interfere with my schemes of geologizing in the Southern states. In the letters, therefore, of introduction furnished to me at Washington, it was particularly requested that information respecting my objects, and facilities of moving speedily from place to place, should be given me, instead of dinners and society. These injunctions were every where kindly and politely complied with. It was my intention, for the sake of getting a correct notion of the low country between the granitic

region and the Atlantic, to examine the cliffs bounding
the Savannah river from its rapids to near its mouth, a
distance, including its windings, of about 250 miles.
After passing a few days at Augusta, where, for the
first time, I saw cotton growing in the fields, I embarked
in a steam-boat employed in the cotton trade, and went
for forty miles down the great river, which usually
flows in a broad alluvial plain, with an average fall of
about one foot per mile, or 250 feet between Augusta
and the sea. Like the Mississippi and all large rivers,
which, in flood season, are densely charged with sediment, the Savannah has its immediate banks higher
than the plain intervening between them and the high
grounds beyond, which usually, however distant from
the river, present a steep cliff or "bluff" towards it.
The low flat alluvial plain, overflowed in great part at
this rainy season, is covered with aquatic trees, and an
ornamental growth of tall canes, some of them reaching
a height of twenty feet, being from one to two inches
in diameter, and with their leaves still green. The
lofty cedar, *now* leafless, towers them, and is remarkable
for the angular bends of the top boughs, and the large
thick roots which swell out near the base.

I landed first at a cliff about 120 feet high, called
Shell Bluff, from the large fossil oysters which are
conspicuous there. About forty miles below Augusta,
at Demery's Ferry, the place where we disembarked,
the waters were so high that we were carried on shore
by two stout negroes. In the absence of the proprietor
to whom I had letters, we were hospitably received by

his overseer, who came down to the river bank, with two led horses, on one of which was a lady's saddle. He conducted us through a beautiful wood, where the verdure of the evergreen oaks, the pines, and hollies, and the mildness of the air, made it difficult for us to believe that it was mid-winter, and that we had been the month before in a region of snow storms and sledges. We crossed two creeks, and after riding several miles reached the house, and were shown into a spacious room, where a great wood fire was kept up constantly on the hearth, and the doors on both sides left open day and night.

Returning home to this hospitable mansion in the dusk of the evening of the day following, I was surprised to see, in a grove of trees near the court-yard of the farm, a large wood-fire blazing on the ground. Over the fire hung three cauldrons, filled, as I afterwards learned, with hog's lard, and three old negro women, in their usual drab-coloured costume, were leaning over the cauldrons, and stirring the lard to clarify it. The red glare of the fire was reflected from their faces, and I need hardly say how much they reminded me of the scene of the witches in Macbeth. Beside them, moving slowly backwards and forwards in a rocking-chair, sat the wife of the overseer, muffled up in a cloak, and suffering from a severe cold, but obliged to watch the old slaves, who are as thoughtless as children, and might spoil the lard if she turned away her head for a few minutes. When I inquired the meaning of this ceremony I was told it was "killing

time," this being the coldest season of the year, and
that since I left the farm in the morning thirty hogs
had been sacrificed by the side of a running stream not
far off. These were destined to serve as winter pro-
visions for the negroes, of whom there were about a
hundred on this plantation. To supply all of them with
food, clothes, and medical attendants, young, old, and
impotent, as well as the able-bodied, is but a portion
of the expense of slave-labour. They must be continually superintended by trustworthy whites, who
might often perform no small part of the task, and
far more effectively, with their own hands.

Resuming our voyage, thirty miles down the river
in another large cotton steam-boat, we were landed
at Stony Bluff, in Georgia, where I wished to examine
the rocks of burr-stone. There was no living being or
habitation in sight. The large steamer vanished in an
instant, sweeping down the swollen river at the rate
of seventeen miles an hour, and it seemed as if we had
been dropped down from a balloon, with our luggage
in the midst of a wilderness. After making a collection
of specimens, I walked about the wood, and found a
lone house, at the door of which a woman was sitting
in a languid state of health. She said she had just
recovered from the fever, or chill; and among other
inquiries, asked when we had last had this complaint.
On being told we had never had it, she said, "I should
like to live in your country, for among the Whites there
is not one in this section of Georgia that has escaped.'
It is true, that consumption, so common in the Northern

states, and so often fatal, is unknown here, but the
universality of the ague makes these low districts in
the Southern states most unenviable dwelling-places.
The best season for a geological tour in this part of
Georgia and South Carolina, east of the mountains,
is from December to April inclusive.

I waited for the return of the owner of the lone
house, and told him I wished to visit the plantation of
Colonel Jones, at Millhaven. He consented to let me
hire his barouche with one horse, telling me I must
send it back the best way I could, after finding my
own way for twelve miles through the pine forests, as
he could spare me no driver. The lanes through the
wood were numerous, and a storm had blown down
so many tall pines across the road, each of which it was
necessary to circumnavigate, that we thought ourselves fortunate when we arrived safe at the destined
haven. My new host added to the kindness and frank-
ness of a Southern planter, a strong love for my fav-
ourite pursuits, and guided me at once to Jackson-
borough, and other neighboring places, best worthy
the attention of a geologist.

We had many rides together through those woods,
there being no underwood to prevent a horse from
galloping freely in every direction. The long-leaved
pines emit a faint odour somewhat resembling that of .
the hyacinth, and their bright-green foliage was finely
brought out against the clear blue sky. The air was
balmy, and unusually warm, even for Georgia in the
first week of January. We saw several butterflies, one

of a bright yellow colour, and bats flying about in the
evening. The croaking of the frog and the chirping
of the cricket were again heard. They had been silent
a few days before, when the air was cooler. The sheep,
which remain out in these woods all the winter, are
now followed by lambs about three weeks old. I saw
many black squirrels here, but only heard of the opos
sum, racoon, bear, and alligator, without seeing any.
A few days ago, an alligator was shot fourteen feet
long, in the act of carrying off a pig; and the sports
men complain to me that they devour their dogs when
they follow the deer, which, on the first alarm, usually
take to the Savannah river.

I frequently observed the holes of the gopher, a kind
of land-tortoise, which burrows in the sand, and is
now hybernating below ground. Four or five inhabit
one hole; their eggs are rather smaller than a hen's.
They are gregarious, and in summer are seen feeding
ten or twelve together on the low shrubs. They are
said to be very strong for their size, and a negro-
woman assured a lady of our party that she was so
light that she might be "toted by a gopher." We also
saw small hillocks, such as are thrown up by our moles,
made by a very singular animal, which they call the
salamander, because, I believe, it is often seen to appear when the woods are burnt. It is not a reptile, but
a species of rat, with pouches on its cheeks.

On quitting Millhaven, instead of continuing my
voyage down the river, I hired a carriage to convey us
to the town of Savannah, a distance of nearly one

hundred miles. Here and there I went down from the high road to examine the river-cliffs, consisting of bright red-coloured loam, red and grey clay, and white sand. One day, on returning from the river, I came suddenly in the wood on some turkey-buzzards feeding on a dead hog. I had often seen since we crossed the Potomac these large black and grey birds soaring at a great height in the air, but I was now surprised to see one of them perch on a stump a few yards from me, and seem perfectly fearless. In our last day's journey, I remarked, for the first time in America, a large flight of rooks, some wheeling about in the air, others perched on trees.

Near the village of Ebenezer we passed over a long causeway, made of logs, which for three quarters of a mile was under water. The tall cedars, and other trees arching over and forming a long aisle, reminded me exactly of the descriptions given of the canals in the Great Dismal Swamp. Some of the myrtles in these wet grounds are very fragrant.

We were pursuing a line of road not much frequented of late, since the establishment of the railway from Augusta to Charleston. Our arrival, therefore, at the inns was usually a surprise, and instead of being welcomed, we were invariably recommended to go on farther. When once admitted, we were made very comfortable, having our meals with the family, and being treated more like guests than customers. On one occasion our driver, to whose brother our carriage and horses belonged, fell in with the son of a neighbor-

ing planter, who reproached him in a friendly manne
for not having come to his house the night before, an
brought us with him. The social equality which pre
vails here arises not so much from the spirit of a re
publican government, as from the fact of the white
constituting an aristocracy, for whom the negroe
work. Had we availed ourselves of letters of intro
duction freely offered to us, we might have passe
from the house of one hospitable planter to another
and heard as little of reckoning at inns as Don Quixot
expected, after his study of the histories of knight
errant.

Jan. 10, 1842.—On the tenth day after leavin
Augusta, we arrived at Savannah, from which tow
I immediately set out on an excursion through a flat
swampy country, resembling a large delta, to Beaul
and the Vernon river, about fifteen miles to the south
east. I went to Heyner's Bridge, on the White Bluf
creek, to see a spot about twelve miles from Savannah
where I had learnt from Dr. Habersham that bones o
the mastodon and other extinct mammalia had bee
discovered. The bed of clay, about six feet thick
containing them, can only be seen at low water, and
descended to it in a boat when the tide was out; an
by the aid of the negroes, obtained the grinder of th
common American mastodon. The stratum enclosin
these and other bones rests immediately on sand con
taining marine shells of living species, and is covere
by the mud of a freshwater swamp, in which tree
grow, and when thrown down by the winds, becom

occasionally imbedded. One of the teeth given to me from this place by Dr. Habersham was ascertained, by Mr. Owen, to be referable to his new genus, Mylodon. Mr. Hamilton Couper afterwards sent me from a similar geological position, farther south in Georgia, near the mouth of the Altamaha, the tooth of a megatherium. It is evident, from his observations and my own, that at a comparatively recent period since the Atlantic was inhabited by the existing species of marine testacea, there was an upheaval and laying dry of the bed of the ocean in this region. The new land supported forests in which the megatherium, mylodon, mastodon, elephant, a species of horse different from the common one, and other quadrupeds, lived, and were occasionally buried in the swamps. There have also been subsidences on the coast, and perhaps, far inland; for in many places near the sea there are signs of the forest having become submerged, the remains of erect trees being seen enveloped in stratified mud and sand; I even suspect that this coast is now sinking down, at a slow and insensible rate, for the sea is encroaching and gaining at many points on the freshwater marshes. Thus at Beauly I found upright stumps of trees of the pine, cedar, and ilex covered with live oysters and barnacles, and exposed at low tide, the deposit in which they were buried having been recently washed away from around them by the waves. I also observed, that the flat country of marshes was bounded on its western or inland side by a steep bank or ancient cliff cut in the sandy tertiary

strata, and there are other inland cliffs of the sam
kind at different heights implying the successive ele
vation above the sea of the whole tertiary region.

On the beach at Beauly I saw numerous foot-track
of racoons and opossums on the sand, which had bee
made during the four hours immediately preceding
or since the ebbing of the tide. Already some of then
were half filled with fine blown sand, showing th
process by which distinct casts may be formed of th
footsteps of animals in a stratum of quartzose sand
stone. I remarked that the tracks of the racoon
could be traced at several points to beds of oysters
on which these animals are said to feed. The negroe
told me, that sometimes a large oyster closes his she
suddenly, and holds the racoon fast by the paw till th
returning tide comes up and drowns him.

The surface of the beach for half a mile was covere
with small round pellets of mud as thick as hail-stones
of the size of currants and peas, and arranged for th
most part in small heaps. These are made by thou
sands of land crabs, which they call fiddlers, becaus
the motion of their claws is compared to the arm of
player on the violin. By the side of each heap wa
a perpendicular hole several inches deep, into whic
when alarmed the crab retreats sideways, sometime
disappearing, but often leaving the larger claw pro
jecting above for want of room. They make thes
holes by rolling the wet sand into pellets, and the
bringing up each ball separately to the surface.

A planter of this country told me it was amusing t

see a flock of turkies driven down for the first time from the interior to feed on the crabs in the marine marshes. They, at first, walk about in a ludicrous state of alarm, expecting their toes to be pinched, but after a time, one bolder than the rest is tempted by hunger to snap up a small fiddler, after which the rest fall to and devour them by thousands. On my way through the woods in this low region near Savannah, I saw some fine magnolias ninety feet high, palmettos six feet high in tufts, and oaks hung with white pendant wreaths, sometimes ten feet long, of the wiry parasitic *Tillandsia usnaeoides*. This climber, which also festoons the woods in South America, much resembles the lichen called in England "old man's beard," but is a phenogamous plant.

In order to see the bed of clay containing the bones of the mastodon at Heyner's Bridge, it was necessary for me to be on the ground by daybreak at low tide. With this in view, I left Savannah in the middle of the night. The owner of the property kindly lent me his black servant as a guide, and I found him provided with a passport, without which no slave can go out after dusk. The exact streets through which he was to pass in his way to me were prescribed, and had he strayed from this route he might have been committed to the guard-house. These and other precautionary regulations, equally irksome to the slaves and their masters, are said to have become necessary after an insurrection brought on by abolitionist missionaries, who are spoken of here in precisely the same tone as incendiaries, or beasts of prey whom it would be meritorious to shoot or hang. In this savage and determined spirit I heard some planters speak who were mild in their manners, and evidently indulgent to their slaves. Nearly half the entire population of this state are of the coloured race, who are said to be as excitable as they are ignorant. Many proprietors live with their wives and children in the midst of the slaves, so that the danger of any popular movement is truly appalling.

The negroes, so far as I have yet seen them, whether in domestic service or on the farms, appear very cheerful and free from care, better fed than a large part of the labouring class of Europe; and, although meanly dressed and often in patched garments, never scantily clothed for the climate. We asked a woman in Georgia, whether she was the slave of the family of our acquaintance. She replied, merrily, "Yes, I belong to them, and they belong to me." She was, in fact, born and brought up on the estate.

On another occasion we were proceeding in a well-appointed carriage with a planter, when we came unexpectedly to a dead halt. Inquiring the cause, the black coachman said he had dropped one of his white gloves on the road, and must drive back and try to find it. He could not recollect within a mile where he had last seen it: we remonstrated, but in vain. As time pressed, the master in despair took off his own gloves, and saying he had a second pair, gave them to him. When our charioteer had deliberately put them on, we started again.

CHAPTER IX

JAN. 13, 1842.—From Savannah we returned to Charleston in a steam-ship, on board of which we found an agreeable party, consisting chiefly of officers of the U. S. army returning from Florida, where they had nearly brought to a close a war of extermination carried on for many years against the Seminole Indians. They gave a lively picture of the hardships they underwent in the swamps and morasses during this inglorious campaign, in the course of which the lives of perhaps as many whites as Seminoles were sacrificed. The war is said to have been provoked by the attacks of the Indians on the new settlers.

When discoursing here on the influence of climate, many accounts were given me of a frost which visited Charleston in February, 1835, so severe that wine was frozen in bottles. The tops of the Pride-of-India tree, of Chinese origin, were killed: all the oranges, of which there were large orchards, were destroyed. Beds of oysters, exposed between high and low water mark, perished in the estuaries, and the effluvia from them was so powerful as to injure the health of the inhabitants.

Several planters attribute the failure of the cotton crop this year (1842) to the unusual size and number of the icebergs, which floated southwards last spring from Hudson's and Baffin's Bays, and may have cooled the sea and checked the early growth of the cotton plant. So numerous and remote are the disturbing causes in meteorology! Forty degrees of latitude intervene between the region where the ice-floes are generated and that where the crops are raised, whose death-warrant they are supposed to have carried with them.

On the banks of the Cooper river, we heard occasionally the melodious and liquid note of the mocking-bird in the woods. It is of a fearless disposition, and approaches very near to the houses. I can well imagine that in summer, when the leaves are out, and the flowers in full splendor, this region must be most beautiful. But it is then that the planters are compelled by the fever and ague to abandon their country seats. It was not so formerly. When the English army was campaigning on the Cooper and Santee rivers in the revolutionary war, they encamped with impunity in places where it would now be death to remain for a few days in the hot season. I inquired what could have caused so great a change and found the phenomenon as much a matter of controversy as the origin of the malaria in Italy. The clearing away of the wood from large spaces is the chief alteration in the physical condition of this region in the course of the last sixty years, whereby the damp and swampy grounds undergo annually the process of being dried up by a burning sun. Marshes which are overflowed by the tide twice in every twenty-four hours near the neighbouring coast, both in South Carolina and Georgia, are perfectly healthy. Dr. Arnold remarks, in his Roman History, that Rome was more healthy before the drainage of the Campagna, and when there was more natural wood in Italy and in northern Europe generally. In the southern States of the Union there are no fevers in the winter, at a season when there is no large extent of damp and boggy soil exposed to a hot sun, and undergoing desiccation.

As there were no inns in that part of South Carolina through which we passed in this short tour, and as we were every where received hospitably by the planters, I had many opportunities of seeing their mode of life, and the condition of the domestic and farm slaves. In some rich houses maize, or Indian corn, and rice were entirely substituted for wheaten bread. The usual style of living is that of English country gentlemen. They have well-appointed carriages and horses, and well-trained black servants. The conversation of the gentlemen turned chiefly on agricultural subjects, shooting, and horse-racing. Several of the mansions were surrounded with deer-parks.

Arriving often at a late hour at our quarters in the evening, we heard the negroes singing loudly and joyously in chorus after their day's work was over. On one estate, about forty black children were brought up daily before the windows of the planter's house, and fed in sight of the family, otherwise, we were told, the old women who have charge of them might, in the absence of the parents, appropriate part of their allowance to themselves. All the slaves have some animal food daily. When they are ill, they sometimes refuse to take medicine except from the hands of the master or mistress; and it is of all tasks the most delicate for the owners to decide when they are really sick, and when only shamming from indolence.

After the accounts I had read of the sufferings of slaves, I was agreeably surprised to find them, in general, so remarkably cheerful and light-hearted. It is true that I saw no gangs working under overseers on sugar-plantations, but out of two millions and a half of slaves in the United States, the larger proportion are engaged in such farming occupations and domestic services as I witnessed in Georgia and South Carolina. I was often for days together with negroes who served me as guides, and found them as talkative and chatty as children, usually boasting of their master's wealth, and their own peculiar merits.

South Carolina is one of the few states where there is a numerical preponderance of slaves. One night, at Charleston, I went to see the guard-house, where there is a strong guard kept constantly in arms, and on the alert. Every citizen is obliged to serve in person, or find a substitute; and the maintenance of such a force, the strict laws against importing books relating to emancipation, and the prohibition to bring back slaves who have been taken by their masters into free states, show that the fears of the owner, whether well-founded or not, are real.

A philanthropist may well be perplexed when he desires to devise some plan of interference which may really promote the true interests of the negro. But the way in which the planters would best consult their own interests appears to me very clear. They should exhibit more patience and courage towards the abolitionists, whose influence and numbers they greatly overrate, and lose no time in educating the slaves, and encouraging private manumission to prepare the way for general emancipation. All seem agreed that the states most ripe for this great reform are Maryland, Virginia, North Carolina, Tennessee, Kentucky, and Missouri. Experience has proved in the northern States that emancipation immediately checks the increase of the coloured population, and causes the relative number of the whites to augment very rapidly. Every year, in proportion as the north-western States fill up, and as the boundary of the new settlers in the west is removed farther and farther, beyond the Mississippi and Missouri, the cheaper and more accessible lands south of the Potomac will offer a more tempting field for colonization to the swarms of New Englanders, who are averse to migrating into slave states. Before this influx of white labourers, the coloured race will give way, and it will require the watchful care of the philanthropist, whether in the north or south, to prevent them from being thrown out of employment and reduced to destitution.

If due exertions be made to cultivate the minds, and protect the rights and privileges of the negroes, and it nevertheless be found that they cannot contend, when free, with white competitors, but are superseded by them, still the cause of humanity will have gained. The coloured people, though their numbers remain stationary, or even diminish, may in the meantime be happier than now, and attain to a higher moral rank. They would, moreover, escape the cruelty and injustice which are the invariable consequences of the exercise of irresponsible power, especially where authority must be sometimes delegated by the planter to agents of inferior education and coarser feelings. And last not least, emancipation would effectually put a stop to the breeding, selling, and exporting of slaves to the sugar-growing States of the South, where, unless the accounts we usually read of slavery be exaggerated and distorted, the life of the negro is shortened by severe toil and suffering.

CHAPTER X

Wilmington, N. C.—Mount Vernon—Return to Philadelphia—Reception of Mr. Dickens.

JAN. 22.—I now turned my course northwards, and, after a short voyage in a steamer from Charleston, landed at Wilmington, in North Carolina. I then went by railway to South Washington, visiting several farms on the banks of the north-east branch of Cape Fear river.

On several of the small plantations here I found the proprietors by no means in a thriving state, evidently losing ground from year to year, and some of them talking of abandoning the exhausted soil, and migrating with their slaves to the south-western States. If, while large numbers of the negroes were thus carried to the South, slavery had been abolished in North Carolina, the black population might ere this have been reduced in numbers considerably, without suffering those privations to which a free competition with white labourers must expose them, wherever great facilities for emigration are not afforded.

A railway train shooting rapidly in the dark through the pine forests of North Carolina has a most singular appearance, resembling a large rocket fired horizontally, with a brilliant stream of revolving sparks extending behind the engine for several hundred yards, each spark being a minute particle of wood, which after issuing from the chimney of the furnace, remains ignited for several seconds in the air. Now and then these fiery particles which are invisible by day, instead of lagging in the rear, find entrance by favour of the wind through the open windows of the car, and, while some burn holes in the traveller's cloak, others make their way into his eyes, causing them to smart most painfully.

From the deck of our steam-boat on the Potomac we saw Mount Vernon, formerly the plantation of General Washington. Instead of exhibiting, like farms in the northern States, a lively picture of progress and improvement, this property was described to me by all as worn out, and of less value now than in the days of its illustrious owner. The bears and wolves, they say, are actually re-entering their ancient haunts, which would scarcely have happened if slavery had been abolished in Virginia.[1]

The air was balmy on the Potomac the last day of January, and the winter had been so mild in the southern States, that we were surprised, on recrossing the Susquehanna at Havre de Grace in Maryland, to see large masses of floating ice brought down from the Appalachian hills, and to feel the air sensibly cooled while we were ferried over the broad river. It struck me as a curious coincidence, and one not entirely accidental, that, precisely in this part of our journey, I once more

saw the low grounds covered with huge boulders, re-
minding me how vast a territory in the South I had
passed over without encountering a single erratic
block. These far transported fragments of rock are
decidedly a northern phenomenon, or belong to the
colder latitudes of the globe, being rare and exceptional
in warmer regions.

Philadelphia, Feb. 1.—The newspapers are filled
with accounts of the enthusiastic reception which Mr.
Charles Dickens is meeting with every where. Such
homage has never been paid to any foreigner since
Lafayette visited the States. The honours may appear
extravagant, but it is in the nature of popular enthu-
siasm to run into excess. I find that several of my
American friends are less disposed than I am to sympathise with the movement, regarding it as more akin to
lion-hunting than hero-worship. They express a doubt
whether Walter Scott, had he visited the U. S., would
have been so much idolised. Perhaps not; for Scott's
poems and romances were less extensively circulated
amongst the millions than the tales of Dickens. There
may be no precedent in Great Britain for a whole people
thus unreservedly indulging their feelings of admiration
for a favorite author; but if so, the Americans deserve
the more credit for obeying their warm impulses. Of
course, many who attend the foreigner's crowded levee
are merely gratifying a vulgar curiosity by staring at
an object of notoriety; but none but a very intelligent
population could be thus carried away to flatter and
applaud a man who has neither rank, wealth, nor power,

who is not a military hero or a celebrated political character, but simply a writer of genius, whose pictures of men and manners, and whose works of fiction, have been here, as in his own country, an inexhaustible source of interest and amusement.

CHAPTER XI

Philadelphia—Financial Crisis—Payments of State Dividends suspended—General Distress and private Losses of the Americans—Debt of Pennsylvania—Public Works—Direct Taxes—Deficient Revenue—Bad Faith and Confiscation—Solvency and Good Faith of the Majority of the States—Confidence of American Capitalists—General Progress of Society, and Prospects of the Republic.

PHILADELPHIA, January to March, 1842.—Wishing to borrow some books at a circulating library, I presented several dollar notes as a deposit. At home there might have been a ringing of coin upon the counter, to ascertain whether it was true or counterfeit, here the shopwoman referred to a small pamphlet, re-edited "semi-monthly," called a "Detector," and containing an interminable list of banks in all parts of the Union, with information as to their present condition, whether solvent or not, and whether paying in specie, and adding a description of "spurious notes." After a slight hesitation, the perplexed librarian shook her head, and declaring her belief that my notes were as good as any others, said, if I would promise to take them back again on my return, and pay her in cash, I might have the volumes.

It often happened that when we offered to buy articles of small value in shops, or fruit in the market, the venders declined to have any dealings with us, unless we paid in specie. They remarked that their change

might in a few days be worth more than our paper.
Many farmers and gardeners are ceasing to bring their
produce to market, although the crops are very abundant, and prices are rising higher and higher, as if the
city was besieged. My American friends, anxious that
I should not be a loser, examined all my dollar notes,
and persuaded me, before I set out on my travels, to
convert them into gold, at a discount of eight per cent.
In less than four weeks after this transaction, there
was a general return to cash payments, and the four
banks by which the greater part of my paper had been
issued, all failed.

A parallel might perhaps be found for a crash of this
kind in the commercial and financial history of England, or at least in some of her colonies, Australia, for
example, where the unbounded facility afforded to a
new country of borrowing the superabundant capital
of an old one, has caused a sudden rise in the value of
lands, houses, and goods, and promoted the maddest
speculations. But an event now occurred of a different
and far more serious nature. One morning we were
told that the Governor of Pennsylvania had come in
great haste from Harrisburg, in consequence of the
stoppage of one of the banks in the city, in which were
lodged the funds intended for the payment of dividends
on state bonds, due in a few days. On this emergency
he endeavoured to persuade other banks to advance the
money, but in vain; such was the general alarm, and
feeling of insecurity. The consequent necessity of a
delay of payment was announced, and many native

holders of stock expressed to me their fears, that although they might obtain the dividend then actually due, it might be long before they received another. At the same time they declared their conviction, that the resources of the State, if well managed, were ample; and that, if it depended on the more affluent merchants of Philadelphia, and the richer portion of the middle class generally, to impose and pay the taxes, the honour of Pennsylvania would not be compromised.

It was painful to witness the ruin and distress occasioned by this last blow, following, as it did, so many previous disasters. Men advanced in years, and retired from active life, after success in business, or at the bar, or after military service, too old to migrate with their families to the West, and begin the world again, are left destitute; many widows and single women have lost their all, and great numbers of the poorer classes are deprived of their savings. An erroneous notion prevails in England that the misery created by these bankruptcies is confined chiefly to foreigners, but, in fact, many of the poorest citizens of Pennsylvania, and of other states, had invested money in these securities. In 1844, or two years after my stay in Philadelphia, The Savings' Bank of New York presented a petition to the legislature at Harrisburg for a resumption of payment of dividends, in which it is stated that their bank then held 300,000 dollars, and had held 800,000, but was obliged to sell 500,000 at a great depreciation, in order to pay the claimants, who were compelled by the distress of the times to withdraw their deposits.

It appears that in the year 1831, when Pennsylvania borrowed a large sum for making railways and canals, she imposed direct taxes for seven years, for the express purpose of regularly paying the interest of her debt. It was hoped, from the experience of New York, that, at the expiration of that term of years, the public works would become sufficiently profitable to render it unnecessary to renew the tax. The inhabitants went on paying until the year 1836, when the government thought itself justified in remitting the burden, on being unexpectedly enriched by several large sums from various sources. In that year they received for granting a charter to the U. S. Bank of Pennsylvania 2,600,000 dollars, and 2,800,000 more for their share of monies which had accumulated in the treasury of the Federal Government, arising out of the sale of public lands, and then divided among the states. It was calculated that these funds would last for three years, and that the public works would by that time yield a revenue sufficient to defray the interest of the sum laid out on executing them.

That the legislature should have seized the first opportunity of relieving their constituents from the direct taxes will astonish no one who has perused the printed paper of the tax-assessor in Pennsylvania, which every one is required to fill up. The necessity of ascertaining the means of persons possessed of small property renders the questions exceedingly minute and inquisitorial. From a variety of others, I extract the following:—"What is the amount of your monies loaned on mortgage, and the debts due to you by solvent debtors?" "What interest do they pay?" "What shares do you hold in any bank or company in any other State?" "How many pleasure carriages do you keep?" "How many watches do you own?—are they gold or silver?" and so forth.

Soon after the ill-judged remission of this tax, a great combination of circumstances led to over-trading, and the most extravagant schemes of money-making. The United States' Bank, during its controversy with President Jackson, had accumulated a large amount of specie, and lent it out most lavishly and imprudently; and when it obtained a new charter from Pennsylvania, it again promoted loans of all kinds, which gave an inordinate stimulus to speculation. Some of the great London banks, at the same time, gave credit to a prodigious amount, often without sufficient caution; and when they were compelled to withdraw this credit suddenly, they had no time to distinguish which of their creditors were worthy of confidence. A great fire in New York, in 1835, had annihilated property to the value of six millions sterling. After the United States' Bank had ceased to be connected with the Federal Government, many other States, besides Pennsylvania, granted charters to banks, which led to an over-issue of notes, and a hot-bed forcing of trade throughout the Union. Then came, in 1839, the miserable expedient of authorizing banks to suspend cash payments, and in 1841, the stoppage of the great U. S. Bank of Pennsylvania, followed by a general panic and financial crisis.

CHAPTER XII

New York City—Residence in New York—Effects on Society of increased Intercourse of distant States—Separation of the Capital and Metropolis—Climate—Lectures for the Working Classes.

NEW YORK, March, 1842.—We spent several weeks at New York, and soon found ourselves at home in the society of persons to some of whom we had letters of introduction from near relatives in England, and others whom we had met at distant places in the course of our tour. So many American citizens migrate from north to south for the sake of the mild winters, or attendance on Congress, or the supreme courts at Washington, or congregate in large watering places during the summer, or have children or brothers in the Far West; everywhere there is so much intercourse, personal and epistolary, between scientific and literary men in remote states, who have often received their university education far from home, that in each new city where we sojourn our American friends and acquaintances seem to know something of each other, and to belong to the same set in society. The territorial extent and political independence of the different States of the Union remind the traveller rather of the distinct nations of Europe than of the different counties of a single kingdom like England; but the population has spread so fast from certain centres, especially from New England, and the facilities of communication by railway and steam-boat are so great, and are always improving so rapidly, that the twenty-six republics of 1842, having a population of seventeen millions, are more united, and belong more thoroughly to one nation than did the thirteen States in 1776, when their numbers were only three millions. In spite of the continual decline of the federal authority, and the occasional conflict of commercial interests between the North and the South, and the violent passions excited by the anti-slavery movement, the old colonial prejudices have been softening down from year to year, the English language, laws, and literature, have pervaded more and more the Dutch, German, and French settlements, and the danger of the dismemberment of the confederacy appears to all reflecting politicians less imminent now than formerly.

I dined with Mr. Astor, now far advanced in years, whose name is well known to the readers of Washington Irving's "Astoria." He informed me that he was about to found a large public library in New York, which I rejoice to hear, as the scientific men and naturalists of this country can rarely afford to purchase expensive European works with numerous illustrations. I often regretted, during my short residence here, that the town of Albany, 150 miles distant, is destined, because it is the capital, to possess the splendid collection of minerals, rocks, and fossils obtained during the late government survey. The surveyors are now employed in arranging these treasures in a museum, which would have been far more useful and more frequently consulted if placed in the midst of this wealthy metropolis, having a population of 300,000 souls. Foreigners, indeed, who have only visited New York for commercial purposes, may imagine that all the inhabitants are exclusively engrossed with trade and money-making; but there is a college here, and many large and flourishing literary and scientific institutions. I received numerous invitations to deliver lectures on geology, but had scarcely time to finish one short course when I was reminded, by the breaking up of winter, that I could resume my operations in the field.[1]

It was now the second week of April, and already the willows on the "Battery" were putting forth their yellowish-green leaves. The air was as warm as in an English summer, although a few days before the ground had been covered with snow. Such sudden changes are trying to many constitutions; and we are told that if we staid a second year in the United States we should feel the influence of the climate, and begin to lose that freshness of colour which marks the newly-arrived Englishman. The greater sallowness of complexion here is attributed to the want of humidity in the air; and we ought to congratulate ourselves that there is no lack of that ingredient in the atmosphere of Great Britain. We continue to be surprised at the clearness of the skies, and the number of fine days and bright star-light nights, on this side of the Atlantic.

At a small New England town in the Taconic hills I

was getting some travelling instructions at the inn, when a carpenter entered who had just finished his day's work, and asked what lecture would be given that evening. The reply was, Mr. N. on the Astronomy of the Middle Ages. He then inquired if it was gratis, and was answered in the negative, the price being twenty-five cents (or one shilling English); upon which he said he should go, and accordingly returned home to dress. It reflects no small credit on the national system of education in New England, that crowds of the labouring classes of both sexes should seek recreation, after the toils of the day are over, in listening to discourses of this kind. Among the most popular subjects of lectures which I saw announced in newspapers were Temperance, a cause which has made great progress of late years among Protestants as well as Catholics, and which began in the U. S. fifteen years before the corresponding movement in Great Britain; Phrenology, to the pretensions of which the Americans lend too credulous an ear; the History of the American Revolution; the Present State and Past History of China; Travels in the Holy Land; Meteorology, and a variety of other topics.

I alluded to some Indians settled near Gayhead, a remnant of the aborigines, who have been protected by the Government of Massachusetts, all sales of land by them to the whites being null and void by law. They make excellent sailors in the whale-fishery of the South Seas, a source of great wealth to the inhabitants of "the Vineyard," and of New Bedford on the main

land.   That occupation, with all its privations and
dangers, seems admirably suited to the bodily constitution and hereditary instinct of a hunter tribe, to whom
steady and continuous labour is irksome and injurious.

The history of the extermination of the aboriginal
Indians of New England is a melancholy tale, especially
after so many successful exertions had been made to
educate and christianize them.   When at Harvard College, a copy of the Bible was shown me by Mr. Jared
Sparks, translated by the missionary Father Elliott
into the Indian tongue.   It is now a dead language, although preached for several generations to crowded
congregations.

CHAPTER XIII

Popular Libraries in New England—Large Sales of Literary Works in the United States—American Universities—Harvard College near Boston—English Universities—Peculiarities of their System.

APRIL 25.—Munificent bequests and donations for public purposes, whether charitable or educational, form a striking feature in the modern history of the United States, and especially of New England. Not only is it common for rich capitalists to leave by will a portion of their fortune towards the endowment of national institutions, but individuals during their lifetime make magnificent grants of money for the same objects. There is here no compulsory law for the equal partition of property among children, as in France, and, on the other hand, no custom of entail or primogeniture, as in England, so that the affluent feel themselves at liberty to share their wealth between their kindred and the public; it being impossible to found a family, and parents having frequently the happiness of seeing all their children well provided for and independent long before their death. I have seen a list of bequests and donations made during the last thirty years, for the benefit of religious, charitable, and literary institutions in the State of Massachusetts alone,

111

and they amounted to no less a sum than six millions of dollars, or more than a million sterling.

There are popular libraries in almost every village of Massachusetts, and a growing taste for the reading of good books is attested by the sale of large editions of such works as Herschel's Natural Philosophy, Washington Irving's Columbus, and Plutarch's Lives. Of each of these from five to twenty thousand copies have been sold. It will seem still more remarkable, that no less than sixteen thousand copies have been purchased of Johnes's Translation of Froissart's Chronicles, illustrated by wood-engravings, and twelve thousand of Liebig's Animal Chemistry. These editions were very cheap, as there was no author's copyright; but it is still more surprising, that about four thousand copies of Prescott's Mexico should have been sold in one year in the U. S. at the price of six dollars, or about twenty-six shillings. When, in addition to these signs of the times, we remember the grants before alluded to, of the New England and other states in behalf of public schools and scientific surveys, we may indulge very sanguine hopes of the future progress of this country towards a high standard of general civilization.

The universities of the United States are annually increasing in number, and their discipline in New England (to which my inquiries on this head were chiefly confined) is very strict; a full staff of professors, with their assistants or tutors, superintending at once the moral conduct and intellectual culture of the students. In Harvard College, Cambridge, near Boston, the best endowed university in the United States, there are thirty-two professors, each assisted by one or more tutors. Many of them are well known in the literary world as authors. Five only of the thirty-two were educated for the pulpit, three of whom are professors of divinity, one of ethics, and one of history. All the students are required to attend divine service in the churches to which they severally belong, but the divinity school for professional education is Unitarian. The proportion of professors to students (about 400 in number) is far greater than that of college tutors in the English universities. The tutors of Harvard College may be compared, in some degree, to our private tutors, except that they are more under the direction of the professors, being selected by them from among the graduates, as the best scholars, and each is specially devoted to some one department of learning. These tutors, from whose number the professors are very commonly chosen, usually teach the freshmen, or first-year students, or prepare pupils for the professors' lectures. Care is also bestowed on the classification of the young men, according to their acquirements, talents, and tastes. To accomplish this object, the student, on entering, may offer to undergo an examination, and, if he succeeds, he may pass at once into the second, third, or fourth year's class, the intermediate steps being dispensed with; he may also choose certain subjects of study, which are regarded as equivalents, or are exchangeable with others. Thus, in the four years of the regular academical course, a competent knowledge of Latin, Greek, and of various branches of mathematics, is exacted from all; but, in regard to other subjects, such as moral philosophy, modern languages, chemistry, mineralogy, and geology, some of them may be substituted for others, at the option of the pupil. There are public examinations at the end of every term for awarding honours or ascertaining the proficiency of students, who, if they have been negligent, are put back into a previous year's class, the period of taking their degree being in that case deferred. Honours are obtainable for almost every subject taught by any professor; but emulation is not relied upon as the chief inducement for study. After passing an examination for the fourth year's class, the student can obtain the degree of Bachelor of Arts, and may enter the divinity, medical, or law schools.

Every inquiry into the present state of the universities in America drew forth from my informants, in return, many questions respecting Oxford and Cambridge. In the first place, then, the mass of students or undergraduates at Oxford is divided into twenty-four separate communities or colleges, very unequal in number, the residents in each varying from 10 in the smaller to about 140 in the larger colleges, and the whole business of educating these separate sections of the youth is restricted to the tutors of the separate colleges. Consequently, two or three individuals, and occasionally a single instructor, may be called upon to give lectures in all the departments of human knowledge embraced in the academical course of four years. If the college is small, there is only occupation and salary sufficient to support one tutor; any attempt, therefore, to sub-divide the different branches of learning and sciences among distinct teachers is abandoned. There is no opportunity for one man to concentrate the powers of his mind on a single department of learning, to en-deavour to enlarge *its* bounds, *and carefully to form* and direct the opinions of his pupil. In a few of the larger colleges, indeed, some rude approach to such a partition is made, so far as to sever the mathematical from the classical studies; but even then the tutors in each division, are often called upon, in the public ex-aminations, to play their part in both departments. Thus, a single instructor gives lectures or examines in the writings of the Greek and Roman historians, phi-losophers, and poets, together with logic, the elements of mathematics, and theology.

For the benefit of my foreign readers, it may be as well to remark, that the scholars to be taught are not boys between the ages of fourteen and eighteen, at which latter age the degree of Bachelor of Arts was very commonly conferred in the olden times at Oxford, but young men between eighteen and twenty-two, who, at the expiration of their academical course, usually quit college and enter at once upon a profession, or into political life.

When we inquire by what kind of training young men can best be prepared, before leaving the university, to enter upon the study or practice of their professions, whether as lawyers, physicians, clergymen, school-

masters, tutors, or legislators, can we assent to the
notion that, by confining instruction to pure mathematics, or the classical writers, more especially if the
latter are not treated in a critical spirit, we shall accomplish this end?   Do not these belong precisely to
the class of subjects in which there is least danger of
the student's going wrong, even if he engages in them
at home and alone?   Should it not be one of our chief
objects to prepare him to form sound opinions in matters connected with moral, political, or physical science?
Here, indeed, he needs the aid of a trustworthy guide,
and director, who shall teach him to weigh evidence,
point out to him the steps by which truth has been
gradually attained in the inductive philosophy, the
caution to be used in collecting facts and drawing con-
clusions, the prejudices which are hostile to a fair inquiry, and who, while his pupil is interested in the
works of the ancients, shall remind him that, as knowledge is progressive, he must avail himself of the latest
acquisitions of his own age, in order to attain views
more comprehensive and correct than those enjoyed
even by predecessors of far superior capacity and
genius.

CHAPTER XIV

Dr. Channing—Agitation in Rhode Island—Armed Convention —Journey to Philadelphia and Baltimore—Harper's Ferry—Passage over the Alleghanies by National Road—Parallel Ridges— Kentucky Farmers—Emigrants.

APRIL 17, 1842.—During my stay at Boston, I was fortunate enough to hear Dr. Channing preach one of the last sermons he delivered from the pulpit. His declining health had prevented him from doing regular duty of late years; but there seemed no reason to anticipate that he would so soon be taken away from a community over which he exerted a great and salutary influence. His sermon was less impressive than I had expected, and fell short of the high conception I had formed of him from his writings; but this I imputed entirely to his want of physical strength, and the weak state of his voice. I had afterwards the pleasure of conversing freely with him at a small dinner party on various subjects in which he was interested; among others, the bearing of geological discoveries, respecting the earth's antiquity and the extinct races of animals, on the Mosaic account of the history of men and the creation. I was struck with the lively interest he took in the political affairs of Rhode Island,—a neighboring state, containing about 110,000 inhabitants, and now

117

convulsed by a revolutionary movement in favor of an extension of the suffrage. The sympathies of Dr. Channing appeared to lean strongly to the popular party, which, in his opinion, had grievances to complain of, however much, by their violent proceedings, they had put themselves in the wrong.

As some alarmists assured me that the railway to Providence, by which I intended to pass southwards in a few days, "was commanded by the cannon of the insurgents," my curiosity was awakened to inquire into this affair, the details of which were not uninstructive, as giving a curious insight into the character of the New England people, and showing their respect for law and order, even when their passions are highly excited. I found that Rhode Island was still, in the year 1842, governed according to a charter granted by Charles II. in the year 1663, no alteration having been made in the qualifications of voters at the period when the sovereignty was transferred from the crown of Great Britian to the freeholders of Rhode Island. Although the State has been flourishing, and is entirely free from debt, a large majority of the people have, for the last forty years, called loudly on the privileged landholders to give up their exclusive right of voting, and to extend the suffrage to all the adult males, in accordance with the system established in all the neighboring States. The dispute turned mainly on a question of a very abstract nature for the comprehension of the multitude, though in reality one of great constitutional importance; namely, whether the change should be made according

to the forms prescribed in the charter of 1663, or might
be effected by the people in its capacity of sovereign,
without regard to any established forms. The latter
method was advocated by the democratic leaders as
most flattering to the people, and with such success
that they organized a formidable association in opposition to the government. Their demands did not dif-
fer very materially from those which the legislature was
willing to concede, except that the democrats claimed
the suffrage, not only for every American-born citizen,
but also for the new-comers, or the settlers of a few
years' standing. Both parties agreed to exclude the
free blacks. At length, as their wishes were not complied with, the "Suffrage Convention" resolved to in-
timidate their opponents by a military enrolment and
drilling, and were soon joined by several companies of
militia.

The governor of Rhode Island was so much alarmed
as to call on the President of the United States to afford him aid, which was declined on the ground that no
overt act of violence had been committed. The insurgents then elected a separate senate and house of
representatives, and one Dorr as governor of the State,
who proceeded to Washington, and had an interview
with the President of the United States and with several
members of congress. Meanwhile military preparations
were making on both sides. A second appeal was made
in vain by the State of Rhode Island for aid from the
federal government at Washington. Meetings of sympathizers were held at New York to co-operate with the

popular party, who had now obtained some pieces of cannon, and attempted to get possession of the arsenal at Providence. On this occasion, however, the State government called out the militia, who mustered in great force, and, after a bloodless affray, the popular party, which had already dwindled down to a few hundreds, deserted their leader, Dorr. This champion made his escape, but was soon after taken, tried for high treason, and condemned to imprisonment. Before the conclusion of this affair the government at Washington signified their readiness to furnish the required troops, but their offer of aid came late, and the assistance was no longer needed.

The firmness of the Rhode Island legislature under the threats of the armed populace at home, and, what was more formidable, of the sympathizers from without, and the respect shown to constitutional forms by the mass of the people in the midst of this excitement, are circumstances highly creditable to the majority of the citizens. It remains to be seen whether an extension of the suffrage, which was afterwards granted, will promote or impede the cause of freedom and good government in this small State.

May 2, 1842.—We now set out on a tour to the valley of the Ohio, and the country west of the Alleghany mountains, taking the railway to Providence, and a steam-boat thence to New York. Afterwards we went to Philadelphia by Amboy, passing through the beautiful straits which separate the mainland of New Jersey from Staten Island. The winding channel is, in parts, only half a mile and even less in width, with many elegant villas and country houses on Staten Island. Its banks are often well wooded, and it resembles a river, or Homer's description of the broad Hellespont, which, as Gibbon observes, the poet had evidently likened to a river, and not to an arm of the sea.

The trees in New England are now only beginning (in the first week of May) to unfold their leaves, after an unusually mild winter. They remain leafless for nearly seven months in the year, although in latitude 42 and 43 N., corresponding geographically to Southern Italy. In New Jersey the scarlet maple is putting forth its young leaves; the horse-chestnuts and lime-trees are in bloom; the lilacs flowering in the gardens, and the Judas tree conspicuous with its purplish pink blossom. The dogwood also abounds in the forests, with such a display of white flowers as to take the place of our hawthorn.

We reached Philadelphia without fatigue in less than twenty-two hours, a distance of 300 miles from Boston, having slept on board the steam-boat between Stonington (Connecticut) and New York. We proceeded from Philadelphia to Baltimore, and from thence ascended the beautiful valley of the Patapsco, for 60 miles to Frederick. At Harper's Ferry, in Virginia, the Potomac, about fifty miles above Washington, is joined by the Shenandoah, a river as large as itself, and after uniting, they issue through a transverse gorge in the mountains. This gorge interested me from its exact resemblance to the Lehigh Gap, in Pennsylvania, by

which the Delaware flows out from the hilly country.
The scenery of Harper's Ferry has been overpraised,
but is very picturesque.

I had hired a carriage at Frederick to carry me to
Harper's Ferry, and thence to Hagerstown, on the main
road across the mountains. When I paid the driver, he
told me that one of my dollar notes was bad, "a mere
personal note." I asked him to explain, when he told
me that he had issued such notes himself. "A friend
of mine at Baltimore," he said, "who kept an oyster
store, once proposed to me to sign twenty-five such
notes, promising that if I would eat out their value in
oysters, he would circulate them. They all passed, and
we never heard of them again." I asked how he reconciled this transaction to his conscience? He replied,
that their currency was in a very unsound state, all
the banks having suspended cash payment, and their
only hope was that matters would soon become so bad
that they must begin to mend. In short, it appeared
that he and his friend had done their best to hasten on
so desirable a crisis.

The next day two Marylanders, one of them the driver
of the stage coach, declared that if the State should impose a property tax, thay would resist payment. As
funds are now wanted to pay the dividends on the
public debt, the open avowal of such opinions in a
country where all have votes, sounded in my ears as
of ominous import.

In our passage over the Alleghanies, we now followed
what is called the National Road to Cumberland and

Frostburg, crossing a great succession of parallel ridges, long and unbroken, with narrow intervening valleys, the whole clothed with wood, chiefly oak. The dogwood, with its white flowers, was very conspicuous. The north-western slopes of the hills were covered with the azalea in full flower, of every shade, from a pale pink to a deep crimson. They are called here the wild honeysuckle. Had not my attention been engrossed with the examination of the geological structure of the numerous parallel chains, the scenery would have been very monotonous, the outline of each long ridge being so even and unbroken, and there being so great a want in this chain of a dominant ridge. There is a remarkable absence of ponds or lakes among these mountains, nor do we see any of those broad, dead flats so common in other chains, especially the Pyrenees, which seem to indicate the place of ancient lakes filled up with sediment. Another peculiarity, also, of a negative kind, is the entire absence of the boulder formation, or drift with transported blocks, which forms so marked a feature in the hills and valleys of New England.

Having one day entered a stage coach in our passage over these mountains, I conversed with two Kentucky farmers returning in high spirits from Baltimore, where they had sold all their mules and cattle for good prices. They were carrying back their money in heavy bags of specie, paper dollars being no longer worthy of trust. They said their crops of grain had been so heavy for several seasons, that it would have cost too much to drag it over the hills to a market 400 miles distant, so

they had "given it legs by turning it into mules."
asked why not horses. They said mules were nearly as
serviceable, and lived longer, coming in for a share o
the longevity of the ass. During several days of travel
ling in public conveyances on this line of route, we me
with persons in all ranks of life, but with no instance o
rude or coarse manners.

Entering a cottage at Frostburg, we talked with the
mother of the family, surrounded by her children and
grandchildren. She appeared prosperous, had left Ire
land forty years before, at the age of seventeen, yet
could not speak of the old country without emotion
saying, "she should die happy could she but once more
see the Cove of Cork." Her children will be more for
tunate, as their early associations are all American.

We passed many waggons of emigrants from Penn
sylvania, of German origin, each encumbered with a
huge, heavy mahogany press, or "schrank," which had
once, perhaps, come from Westphalia. These antique
pieces of furniture might well contain the penates of
these poor people, or be themselves their household
gods, as they seem to be as religiously preserved. Our
companions, the two farmers from Kentucky before
mentioned, shook their heads, remarking, "that most
of them would go back again to Pennsylvania, after
spending all their money in the West; for the old people
will pine for their former homes, and persuade the
younger ones to return with them.

I found some of the iron mines near Frostburg in a
bankrupt state, and met a long train of luggage wag-

gons conveying the familes of the work-people to new settlements in the West. The disappointed speculators are clamouring for a tariff to protect their trade against English competition. When I urged the usual arguments in favour of free trade, I was amused to perceive how the class interests of my new companions had overcome the usual love of equality, which displays itself in the citizens of the United States. One of the superintendents of the mines expressed surprise that I should have gone through so many States, and not grown tired of the dull mediocrity of income which mere land under the custom of equal division among children produced! "Why limit our civilization and refinement to small farmers, who expend their surplus gains in tobacco and lawsuits, and can never make ample fortunes, such as spring from manufacturing and commercial industry?"

CHAPTER XV

*Alleghany Mountains—Union—Horizontal Coal Formations—
Brownsville on the Monongahela—Facilities of Working Coal—
Navigable Rivers—Great Future Resources of the Country—Pittsburg—Fossil Indian Corn—Indian Mounds near Wheeling—General Harrison on their high Antiquity—Dr. Morton on the aboriginal Indians—Remarks on the Civilization of the Mexicans and
other Tribes—Marietta—New Settlements—Cincinnati.*

AFTER leaving the small mining village of Frostburg,
which is about 1500 feet above the level of the sea, we
continued to ascend and descend a succession of steep
ridges till we came to the summit level, where the climate was sensibly colder, and the oaks and other trees
still leafless.   At Smithfield we crossed a river flowing
westward, or toward the Monongahela and Gulf of
Mexico, and soon afterwards passed the grave of General Braddock, and followed the line of his disastrous
march toward Fort Duquesne, now Pittsburg.

At length we reached Laurel Hill, so called from its
rhododendrons, the last of the great parallel ridges of
the Alleghanies.   From this height we looked down
upon a splendid prospect, the low undulating country
to the west, appearing spread out far and wide before
us, and glowing with the rays of the setting sun   At
our feet lay the small town of Union, its site being

126

marked by a thin cloud of smoke, which pleased us by recalling to our minds a familiar feature in the English landscape, not seen in our tour through the regions where they burn anthracite, to the east of the Alleghanies.

After enjoying the view for some time we began to descend rapidly, and at every step saw the forest, so leafless and wintry a few hours before, recover its foliage, till the trees and the climate spoke again of spring. I had passed several times over the Pyrenees and the Alps, and witnessed the changes of vegetation between the opposite flanks, or between the summits and base of those mountains; but this was the first time I had crossed a great natural barrier, and found on the other side people speaking the same language, and having precisely the same laws and political institutions.

At the town of Union, which may be said to lie at the western foot of the mountains, I had an opportunity of seeing coal exposed to view in an open quarry of building stone. The coal seam was three and a half feet thick, with an intervening layer, as usual, between it and the freestone of dark slate or shale, four feet thick. When traced farther, the shale thinned out gradually, and in a neighbouring quarry, about thirty yards distant, it gave place to the yellow micaceous sandstone, which then formed the roof of the coal. These sandstone roofs are comparatively rare in America, as in Europe.

From Union, we went to Brownsville on the Monongahela, a large tributary of the Ohio, where the country consists of coal measures, like those at Union, both evidently belonging to the same series as those more bent and curved beds at Frostburg. I was truly astonished, now that I had entered the hydrographical basin of the Ohio, at beholding the richness of the seams of coal, which appear everywhere on the flanks of the hills and at the bottom of the valleys, and which are accessible in a degree I never witnessed elsewhere. The time has not yet arrived, the soil being still densely covered with the primeval forest, and manufacturing industry in its infancy, when the full value of this inexhaustible supply of cheap fuel can be appreciated; but the resources which it will one day afford to a region capable, by its agricultural produce alone, of supporting a large population, are truly magnificent. In order to estimate the natural advantages of such a region, we must reflect how three great navigable rivers, such as the Monongahela, Alleghany, and Ohio, intersect it, and lay open on their banks the level seams of coal. I found at Brownsville a bed ten feet thick of good bituminous coal, commonly called the Pittsburg seam, breaking out in the river cliffs near the water's edge. Horizontal galleries may be driven everywhere at very slight expense, and so worked as to drain themselves, while the cars, laden with coal and attached to each other, glide down on a railway, so as to deliver their burdens into barges moored to the river's bank. The same seam is seen at a distance, on the right bank, and may be followed the whole way to Pittsburg, fifty miles distant. As it is nearly horizontal, while the river descends, it crops out at a continually increasing, but never at an inconvenient, height above the Monongahela. Below the great bed of coal at Brownsville is a fire-clay eighteen inches thick, and below this, several beds of limestone, below which again are other seams. Almost every proprietor can open a coal-pit on his own land, and, the stratification being very regular, they may calculate with precision the depth at which the coal may be won.

So great are the facilities for procuring this excellent fuel, that already it is found profitable to convey it in flat-bottomed boats for the use of steamships at New Orleans, 1100 miles distant, in spite of the dense forests bordering the intermediate river-plains, where timber may be obtained at the cost of felling it. But no idea can be formed of the importance of these American coal-seams, until we reflect on the prodigious area over which they are continuous. The boundaries of the Pittsburg seam have been determined with accuracy by the Professors Rogers in Pennsylvania, Virginia, and Ohio, and they have found the elliptical area which it occupies to be 225 miles in its longest diameter, while its maximum breadth is about one hundred miles, its superficial extent being about fourteen thousand square miles. While alluding to the vast area of these carboniferous formations in the United States, so rich in productive coal, I may call attention to the Illinois coal-field. That coal-field, comprehending parts of Illinois, Indiana and Kentucky, is not much inferior in dimensions to the whole of England, and consists of horizontal strata, with numerous rich seams of bituminous coal.

May 15, 1842.—We embarked at Brownsville for Pittsburg in a long narrow steamer, which drew only eighteen inches of water, and had a single paddle behind like the overshot wheel of a mill. It threw up a shower of spray like a fountain, which had a picturesque effect. The iron works of the machinery and the furnace were all exposed to view, and the engineers were on deck in a place cooled by the free circulation of air.

The wooded hills rise to the height of from 300 to 450 feet above the river between Brownsville and Pittsburg. The latter place is situated at the junction of the Alleghany and Monongahela rivers, which after their union form the Ohio. It is a most flourishing town, and we counted twenty-two large steamboats anchored off the wharfs. From the summit of the hill, 460 feet high, on the left bank of the Monongahela, we had a fine view of Pittsburg, partially concealed by the smoke of its numerous factories. A great many fine bridges span the two broad rivers above their junction. In the same hill I saw a fine section of the horizontal coal-measures. Far below the principal seam, and near the level of the river, there is a bed of coal a few inches thick, resting on clay.

The steamboats on the Ohio cannot be depended upon for punctual departure at the appointed hour like those of the Hudson or Delaware. I therefore took places in a coach for Wheeling, and crossed a low and nearly level country, where I was struck with the absence of drift and boulders, so common in the north. The carboniferous strata were exposed on the banks of every small streamlet, and not concealed by any superficial covering. On reaching one of those innumerable towns to which, as if for the sake of confusion, the name of Washington has been given, I received the agreeable intelligence that, instead of travelling to Wheeling before sunset, I must wait till another mail came up in the middle of the night. I was very indignant at this breach of promise, but was soon appeased by the good-natured landlord and postmaster, who addressed me by the conciliatory appellation of "Major," and assured me that the new post-office regulation was as inconvenient to him as it could possibly be to us.

The next day we embarked at Wheeling on the Ohio for Marietta. I had been requested by my geological friends, when at Philadelphia, to make inquiries respecting some Indian corn said to have been found fossil at some depth in a stratified deposit near Fish Creek, a tributary of the Ohio, and presumed to be of high antiquity. A proprietor who had resided twenty-six years near the spot, assured me that the corn occurred in an island in the river, at the depth of no more than two feet below the surface of the alluvial soil. It consisted of parched corn, such as the Indians often buried when alarmed, and in the present year the Ohio had risen so high as to inundate the very spot, and throw down several layers of mud upon the site of the corn.

Five miles below Wheeling, on the left bank of the Ohio, is a terrace of stratified sand and gravel, having its surface about seventy-five feet above the Ohio. On this terrace is seen a large Indian mound. On arriving at Marietta, I learnt from Dr. Hildreth that skeletons had been found in it at various depths, together with pipe-heads and other ornaments. Their workmanship implies a more advanced state of the arts than that attained by the rude Indians who inhabited this fertile valley when it was first discovered by the white man. There are many other similar mounds in the valleys of the Ohio and its tributaries, but no tradition concerning their origin. One of these, near Marietta, in which human bones were dug up, must be more than eight centuries old, for Dr. Hildreth counted eight hundred rings of annual growth in a tree which grew upon it. But, however high may be the historical antiquity of the mounds, they stand on alluvial terraces which are evidently of a very modern geological date. In America, as in Europe, the oldest monuments of human labour are as things of yesterday in comparison with the effects of physical causes which were in operation after the existing continents had acquired the leading features of hill and valley, river and lake, which now belong to them. Dr. Locke of Cincinnati has shown that one of the earth-works, enclosing about one hundred acres on the great Miami, although nearly entire, has been overflowed in a few places, and partially obliterated. He infers from this and other facts, that these mounds extending to high-water mark, and liable to be occasionally submerged, were constructed when the streams had already reached their present levels, or, in other words, their channels have not been deepened in the last 1000 or 2000 years.

The arguments for assigning a very remote period to the Indian antiquities above alluded to, have been stated with great force and clearness by General Harrison, late President of the United States, who was practically versed in woodcraft, and all that relates to the clearing of new lands. In his essay on the aborigines of the Ohio valley, he states that some of these earthworks are not mere mounds, but extensive lines of embankment, varying from a few feet to ninety feet in altitude, and enclosing areas of from one to several hundred acres.

"Their sites," he says, "present precisely the same appearance as the circumjacent forest. You find on them all that beautiful variety of trees which give such unrivalled richness to our forests. This is particularly the case on the fifteen acres included within the walls of the work at the mouth of the great Miami, and the relative proportions of the different kinds of timber are about the same."

He then goes on to observe that if you cut down the wood on any piece of wild land, and abandon it to nature, the trees do not grow up as before, but one or two, or at most three species get possession of the whole ground, such for example as the yellow locust, or the black and white walnut. The process by which the forest recovers its original state is extremely slow.

"On a farm of my own," says he, "at the end of fifty years, so little progress had been made, as to show that ten times that period would be necessary to effect its complete assimilation. When those kinds of timber which first establish themselves have for a long time remained undisputed masters of the soil, they at length die by disease, or are thinned by the lightning or tempest. The soil has no longer a preference for them, and by a natural rotation of crops other species succeed, till at length the more homogeneous growth ceases, and the denuded tract is again clothed with a variety of wood." As the sites of the earthworks command extensive views, it is reasonable to infer that no trees were suffered by the Indians to spring up upon and near the mounds, from the state of the surrounding forest, General Harrison concludes that several generations of trees had succeeded each other, before the present trees began to grow, and that the mounds were probably as ancient at least as the Christian era. The rich valley of the Ohio, when first discovered by Europeans, was thinly peopled by rude tribes of Indian hunters. In what manner, then, could they have conquered and driven out that more civilized race which evidently preceded them? Harrison suggests that a great flood, like those which occurred in 1793 and 1832 after heavy rain, when the Ohio was unusually blocked up with ice, may have swept off Indian towns and villages, and caused the terrified occupants to remove. The flood would be construed by their superstition into a warning from heaven to seek a residence upon some smaller streams; and before the remembrance of this fearful calamity had been effaced from their imaginations, the deserted region would, from its great fertility, become the usual resort of game. It would then be a common hunting ground for the hostile tribes of the north and south, and consequently a great arena for battle. In this state it continued when first visited by the whites.

Dr. Morton, in his luminous and philosophical essay on the aboriginal race of America, seems to have proved that all the different tribes, except the Esquimaux, are of one race, and that this race is peculiar and distinct from all others. The physical characteristics of the Fuegians, the Indians of the tropical plains, those of the Rocky Mountains, and of the great valley of the Mississippi, are the same, not only in regard to feature and external lineaments, but also in osteological structure. After comparing nearly 400 crania derived from tribes inhabiting almost every region of both Americas, Dr. Morton has found the same peculiar shape pervading all, " the squared or rounded head, the flattened or vertical occiput, the high cheek bones, the ponderous maxillae, the large quadrangular orbits, and the low receding forehead." The oldest skulls from the cemeteries of Peru, the tombs of Mexico, or the mounds of the Mississippi and Ohio, agree with each other, and are of the same type as the heads of the most savage existing tribes. If we next turn to their arts and inventions, we find that a canoe excavated from a single log was the principal vessel in use throughout the New

World at the period of its discovery, the same primitive model existing among the Fuegians, the predatory Caribs, and the more advanced Mexicans and Peruvians.

But although the various tribes remained in general as stationary in all matters requiring intellectual effort, as in their nautical contrivances, we behold with surprise certain points of which Mexico was the most remarkable, where an indigenous and peculiar civilization had been developed, and had reached a high degree of perfection. However much we may admire their architecture, their picture-writing, and historical records, it is their astronomical science in particular, as Mr. Prescott observes, which was disproportioned to their advancement in other walks of civilization. They had fixed the true length of the tropical [1] year with a precision unknown to the great philosophers of antiquity, which could only be the result of a long series of nice and patient observations. By intercalating a certain number of days into the year at the expiration of every fifty-two years, they had even anticipated the Gregorian reform, so that their calendar at the time of the conquest was more correct than that of the Europeans. To ascribe the civilization of the Toltecs to an Asiatic origin, while it is admitted that there was no correspondence or relationship between their language and that of any known Asiatic nation, appears to me a baseless hypothesis, however true it may be that the aboriginal Americans had in the course of ages derived some hints from foreign sources. They could only have taken advantage of such aid, conjectural as it is, and without proof, if they were already in a highly progressive state; and if such assistance be deemed sufficient to invalidate their title to an independent civilization, no race of mankind can ever make good their claim to such an honor.

If, then, a large continent can be inhabited by hundreds of tribes, all belonging to the same race, and nearly all remaining for centuries in a state of apparently hopeless barbarism, while two or three of them make a start in their social condition, and in the arts and sciences; if these same nations, when brought into contact with Europeans, relapse and retrograde until they are scarcely distinguishable in intellectual rank from the rude hunter tribes descended from a common stock; what caution ought we to observe when speculating on the inherent capacities of any other great member of the human family? The negro, for example, may have remained stationary in all hitherto explored parts of the African continent, and may even have become more barbarous when brought within the influence of the white man, and yet may possess within his bosom the germ of a civilization as active and refined as that of the golden age of Tezcuco.

We were fortunate, when at Pomeroy, to fall in with some New England settlers, who were nearly related to several of our most valued friends at Boston. Their descriptions of what they had gone through since they first founded this flourishing colony in the wilderness, reminded us of that entertaining volume recently published in the United States, called "A New Home: Who'll Follow?" It is not the trees and their rank growth on the uncleared land, nor the wild animals, which are the most uncongenial *neighbours* to persons of superior education and refinement in a new settlement. To enjoy facilities, therefore, of communicating rapidly with the civilized Eastern States by founding their new town on the banks of a great navigable river, or close to some main road in the interior, is a privilege truly enviable. I remember wondering, when I first read Homer's graphic sketch of the advantages of wealth, that he should have placed his rich man's mansion on the road side.

To an Englishman, the poet's notion seemed very unaristocratic, for we are almost irresistibly reminded of the large sums which an English country gentleman would expend in order to remove the high road to a respectful distance. Probably the present condition of Ohio, rather than that of a county of parks and mansions like Surrey, was the model most frequently present to the minds of the migratory Greeks of the Homeric age.

From Pomeroy, a large steamboat carried us more than 200 miles in about fifteen hours, down the broad, winding stream, past many a well-wooded island, to Cincinnati, where we were struck with the appearance of commercial activity, the numerous wharfs and steam boats, the wide streets and handsome buildings.[1]

CHAPTER XVII

Excursion to the Swamps of Big Bone Lick, in Kentucky—Noble Forest—Salt Springs—Buffalo Trails—Numerous Bones of Extinct Animals

Two days after I reached Cincinnati, I set out, in company with two naturalists of that city, Mr. Buchanan and Mr. J. G. Anthony, who kindly offered to be my guides, in an excursion to a place of great geological celebrity in the neighbouring State of Kentucky, called Big Bone Lick, where the bones of mastodons and many other extinct quadrupeds have been dug up in extraordinary abundance. Having crossed the river from Cincinnati, we passed through a forest far more magnificent for the size and variety of its trees than any we had before seen. The tulip-tree, the buckeye, a kind of horse-chestnut, the shagbark hickory, the beech, the oak, the elm, the chestnut, the locust-tree, the sugar-maple, and the willow, were in perfection but no coniferous trees,—none of the long-leaved pines of the Southern Atlantic border, nor the cypress, cedar, and hemlock of other States. These forests, where there is no undergrowth, are called "wood pastures." Originally the cane covered the ground, but when it was eaten down by the cattle, no new crop could get up, and it was replaced by grass alone

Big Bone Lick is distant from Cincinnati about twenty-three miles in a S. W. direction. The intervening country is composed of the blue argillaceous limestone and marl, the beds of which are nearly horizontal and form flat table-lands intersected by valleys of moderate depth. In one of these, watered by the Big Bone Creek, occur the boggy grounds and springs called Licks. The term Lick is applied throughout North America to those marshy swamps where saline springs break out, and which are frequented by deer, buffalo, and other wild animals for the sake of the salt, whether dissolved in the water, or thrown down by evaporation in the summer season, so as to encrust the surface of the marsh. Cattle and wild beasts devour this encrustation greedily, and burrow into the clay impregnated with salt, in order to lick the mud. Bartram, the botanist, tells us, that in his time (1790) he visited Buffalo Lick in Georgia, forming part of a cane swamp, in which the head branches of the Ogeechee River take their rise. The lick consisted of "white-coloured tenacious fattish clay, which all kinds of cattle lick into great hollows, pursuing the delicious vein." "I could discover nothing saline in its taste, but an insipid sweetness. Horned cattle, horses, and deer are immoderately fond of it."

The celebrated bog of Kentucky is situated in a nearly level plain, in a valley bounded by gentle slopes, which lead up to the table-lands before mentioned. The general course of the meandering stream which flows through the plain, is from east to west. There are two springs on the southern or left bank, rising from marshes, and two on the opposite bank, the most western of which, called the Gum Lick, is at the point where a small tributary joins the principal stream. The quaking bogs on this side are now more than fifteen acres in extent, but all the marshes were formerly larger before the surrounding forest was partially cleared away. The removal of tall trees has allowed the sun's rays to penetrate freely to the soil, and dry up part of the morass.

Within the memory of persons now living, the wild bisons or buffaloes crowded to these springs, but they have retreated for many years, and are now as unknown to the inhabitants as the mastodon itself. Mr. Phinnel, the proprietor of the land, called our attention to two buffalo paths or trails still extant in the woods here, both leading directly to the springs. One of these in particular, which first strikes off in a northerly direction from the Gum Lick, is afterwards traced eastward through the forest for several miles. It was three or four yards wide, only partially overgrown with grass, and, sixty years ago, was as bare, hard, and well trodden as a high road.

The bog in the spots where the salt springs rise is so soft, that a man may force a pole down into it many yards perpendicularly. It may readily be supposed, therefore, that horses, cows, and other quadrupeds, are now occasionally lost here; and that a much greater number of wild animals were mired formerly. It is well known that, during great droughts in the Pampas of South America, the horses, cattle, and deer throng to the rivers in such numbers that the foremost of the crowd are pushed into the stream by the pressure of others behind, and are sometimes carried away by thousands and drowned. In their eagerness to drink the saline waters and lick the salt, the heavy mastodons and elephants seem in like manner to have pressed upon each other, and sunk in these soft quagmires of Kentucky.

The greater proportion both of the entire skeletons of extinct animals, and the separate bones, have been taken up from black mud, about twelve feet below the level of the creek. It is supposed that the bones of mastodons found here could not have belonged to less than one hundred distinct individuals, those of the fossil elephant to twenty, besides which, a few bones of a stag, horse, megalonyx, and bison, are stated to have been obtained. Whether the common bison, the remains of which I saw in great numbers in a superficial stratum recently cut open in the river's bank, has ever been seen in such a situation as to prove it to have been contemporaneous with the extinct mastodon, I was unable to ascertain. In regard to the horse, it may probably have differed from our *Equus caballus* [1] as much as the zebra or wild ass, in the same manner as that found at Newberne in North Carolina appears to have done. The greatest depth of the black mud has not been ascertained; it is composed chiefly of clay, with a mixture of calcareous matter and sand, and contains 5 parts in 100 of sulphate of lime, with some animal matter. Layers of gravel occur in the midst of it at various depths. In some places it rests upon the blue limestone. The only teeth which I myself procured from collectors on the spot, besides those of the buffalo, were recognized by Mr. Owen as belonging to extremely young mastodons. From the place where they were found, and the rolled state of some of the accompanying bones, I suspected that they had been washed out of the soil of the bogs above by the river, which often changes its course after floods.

It is impossible to view this plain, without at once concluding that it has remained unchanged in all its principal features from the period when the extinct quadrupeds inhabited the banks of the Ohio and its tributaries. But one phenomenon perplexed us much, and for a time seemed quite unintelligible. On parts of the boggy grounds, a superficial covering of yellow loam was incumbent on the dark-coloured mud, containing the fossil bones. This partial covering of yellow sandy clay was at some points no less than fifteen or twenty feet thick. Mr. Bullock passed through it when he dug for fossil remains on the left bank of the creek, and he came down to the boggy grounds with bones below. We first resorted to the hypothesis that the valley might have been dammed up by a temporary barrier, and converted into a lake; but we afterwards learnt, that although the Ohio is seven miles distant by the windings of the creek, there being a slight descent the whole way, yet that great river has been known to rise so high as to flow up the valley of Big Bone Creek, and so late as 1824, to enter the second story of a house built near the springs. The level of the Licks above the Ohio is about fifty feet, the distance in a straight line being only three miles. At Cincinnati the river has been known to rise sixty feet above its summer level, and in the course of ages it may occasionally have risen higher. It may be unnecessary, therefore, to refer to the general subsidence before alluded to, in order to account for the patches of superficial silt last described.

After spending the day in exploring the Licks, we were hospitably received at the house of a Kentucky proprietor a few miles distant, whose zeal for farming and introducing cattle of the "true Durham breed," had not prevented him from cultivating a beautiful flower garden. We were regaled the next morning with an excellent dish of broiled squirrels. There are seasons when the grey squirrel swarms here in such numbers, as to strip the trees of their foliage, and the sportsmen revenge themselves after the manner of the Hottentots, when they eat the locusts which have consumed every green thing in Southern Africa.

We then returned by another route through the splendid forest, and re-crossed the Ohio. The weather was cool, and we saw no fire-flies, although I had seen many a few days before, sparkling as they flitted over the marshy grounds bordering the Ohio, in my excursion up the river to Rockville.

CHAPTER XVIII

Cincinnati—Journey across Ohio to Cleveland—New Clearings—Rapid Progress of the State since the year 1800—Increase of Population in the United States—Political Discussions—Stump Oratory—Relative Value of Labor and Land.

MAY 29.—We left Cincinnati for Cleveland on Lake Erie, a distance of 250 miles, and our line of route took us through the centre of the State of Ohio, by Springfield, Mount Vernon, and Wooster, at all which places we slept, reaching Cleveland on the fifth day.

In our passage through Ohio, we took advantage of public coaches only when they offered themselves in the day-time, and always found good private carriages for the rest of the way. If some writers, who have recently travelled in this part of America, found the fatigue of the journey excessive, it must have arisen from their practice of pushing on day and night over roads which are in some places really dangerous in the dark. On our reaching a steep hill north of Mount Vernon, a fellow-passenger pointed out to me a spot where the coach had been lately upset in the night. He said that in the course of the last three years he had been overturned thirteen times between Cincinnati and Cleveland, but being an inside passenger had escaped without serious injury.

In passing from the southern to the northern frontier of Ohio, we left a handsome and populous city and
fine roads, and found the towns grow smaller and the
high roads rougher, as we advanced. When more than
half way across the State, and after leaving Mount
Vernon, we saw continually new clearings, where the
felling, girdling, and burning of trees was going on, and
where oats were growing amidst the blackened stumps
on land which had never been ploughed, but only
broken up with the harrow. The carriage was then
jolted for a short space over a corduroy road, constructed of trunks of trees laid side by side, while the
hot air of burning timber made us impatient of the slow
pace of our carriage. We then lost sight for many
leagues of all human habitations, except here and there
some empty wooden building, on which "Mover's
House" was inscribed in large letters. Here we were
told a family of emigrants might pass the night on payment of a small sum. At last the road again improved,
and we came to the termination of the table land of
Ohio, at a distance of about sixteen miles from Lake
Erie. From this point on the summit of Stony Hill we
saw at our feet a broad and level plain covered with
wood; and beyond, in the horizon, Lake Erie, extending far and wide like the ocean. We then began our
descent, and in about three hours reached Cleveland.

The changes in the conditions of the country which
we had witnessed are illustrations of the course of events
which has marked the progress of civilization in this
State, which first began in the south, and spread from

the banks of the Ohio. At a later period, when the great Erie canal was finished, which opened a free commercial intercourse with the river Hudson, New York, and the Atlantic, the northern frontier began to acquire wealth and an increase of inhabitants. Ports were founded on the lake, and grew in a few years with almost unparalleled rapidity. The forest then yielded to the axe in a new direction, and settlers migrated from north to south, leaving still a central wilderness between the Ohio and Lake Erie. This forest might have proved for many generations a serious obstacle to the progress of the State, had not the law wisely provided that all non-resident holders of waste lands should be compelled to pay their full share of taxes laid on by the inhabitants of the surrounding districts for new schools and roads. If an absentee is in arrear, the sheriff seizes a portion of his ground contiguous to a town or village, puts it up for auction, and thus discharges the debt, so that it is impossible for a speculator, indifferent to the local interests of a district, to wait year after year, until he is induced by a great bribe to part with his lands, all ready communication between neighbouring and highly cultivated regions being in the meantime cut off.

Ohio was a wilderness exclusively occupied by the Indians, until near the close of the last century. In 1800 its population amounted to 45,365, in the next ten years it had increased fivefold, and in the ten which followed it again more than doubled. In 1840 it had reached 1,600,000 souls, all free, and almost without any admixture of the coloured race. In this short interval the forest had been transformed into a land of steamboats, canals, and flourishing towns; and would have been still more populous had not thousands of its new settlers migrated still farther west to Indiana and Illinois. A portion of the public works which accelerated this marvellous prosperity, were executed with foreign capital, but the interest of the whole has been punctually paid by direct taxes. There is no other example in history, either in the old or new world, of so sudden a rise of a large country to opulence and power. The State contains nearly as wide an extent of arable land as England, all of moderate elevation, so rich in its alluvial plains as to be cropped thirty or forty years without manure, having abundance of fine timber, a temperate climate, many large navigable rivers, a ready communication through Lake Erie with the north and east, and by the Ohio with the south and west, and, lastly, abundance of coal in its eastern counties.

I am informed that, in the beginning of the present year (1842), the foremost bands of emigrants have reached the Platte River, a tributary of the Missouri. This point is said to be only half way between the Atlantic and the Rocky Mountains, and the country beyond the present frontier is as fertile as that already occupied. De Tocqueville calculated that along the borders of the United States, from Lake Superior to the Gulf of Mexico, extending a distance of more than 1200 miles as the bird flies, the whites advance every year at a mean rate of seventeen miles; and he truly observes that there is a grandeur and solemnity in this gradual and continuous march of the European race towards the Rocky Mountains. He compares it to "a deluge of men rising unabatedly, and daily driven onwards by the hand of God."

When conversing with a New England friend on the progress of American population, I was surprised to learn, as a statistical fact, that there are more whites now living in North America than all that have died there since the days of Columbus. It seems probable, moreover, that the same remark may hold true for fifty years to come. The census has been carefully taken in the United States since the year 1800, and it appears that the ratio of increase was 35 per cent. for the first decennial period, and that it gradually diminished to about 32 per cent. in the last. From these data, Professor Tucker estimated that, in the year 1850, the population will amount in round numbers to 22 millions, in 1860 to 29 millions, in 1870 to 38 millions, in 1880 to 50 millions, in 1890 to 63 millions, and in 1900 to 80 millions.[1]

The territory of the United States is said to amount to one-tenth, or at the utmost to one-eighth of that colonised by Spain on the American continent. Yet in all these vast regions conquered by Cortez and Pizarro, there are considerably less than two millions of people of European blood, so that they scarcely exceed in number the population acquired in about half a century in Ohio, and fall far short of it in wealth and civilization.

CHAPTER XIX

Cleveland—Fredonia, streets lighted with natural Gas—Falls of Niagara—Burning Spring—Passing behind the Falls—Daguerreotype of the Falls.

June 5.—Sailed in a steamboat to Fredonia, a town of 1200 inhabitants, with neat white houses, and six churches. The streets are lighted up with natural gas which bubbles up out of the ground, and is received into a gasometer, which I visited. This gas consists of carburetted hydrogen,[1] and issues from a black bituminous slate. The lighthouse-keeper at Fredonia told me that, near the shore, at a considerable distance from the gasometer, he bored a hole through the black slate, and the gas soon collected in sufficient quantity to explode, when ignited.

We next reached Buffalo, and found so many new buildings erected since the preceding autumn, and new shops opened, that we were amazed at the progress of things, at a time when all are complaining of the unprecedented state of depression under which the commerce and industry of the country are suffering.

At the Falls of Niagara, where we next spent a week, residing in a hotel on the Canada side, I resumed my geological explorations of last summer. Every part of the scenery, from Grand Island above the falls

150

to the Ferry at Queenstown, seven miles below, deserves to be studied at leisure.

We visited the "burning spring" at the edge of the river above the rapids, where carburetted hydrogen, or, in the modern chemical phraseology, a light hydrocarbon, similar to that before mentioned at Fredonia, rises from beneath the water out of the limestone rock. The invisible gas makes its way in countless bubbles through the clear transparent waters of the Niagara. On the application of a lighted candle, it takes fire, and plays about with a lambent flickering flame, which seldom touches the water, the gas being at first too pure to be inflammable, and only obtaining sufficient oxygen after mingling with the atmosphere at the height of several inches above the surface of the stream.

At noon, on a hot summer's day, we were tempted, contrary to my previous resolution, to perform the exploit of passing under the great sheet of water between the precipice and the Horse-shoe Fall. We were in some degree rewarded for this feat by the singularity of the scene, and the occasional openings in the curtain of white foam and arch of green water, which afforded momentary glimpses of the woody ravine and river below, fortunately for us lighted up most brilliantly by a midday sun. We had only one guide, which is barely sufficient for safety when there are *two* persons, for a stranger requires support when he loses his breath by the violent gusts of wind dashing the spray and water in his face. If he turns round to recover, the blast often changes in an instant, and blows as impetuously against him in the opposite direction.

The Falls, though continually in motion, have all the effect of a fixed and unvarying feature in the landscape, like the two magnificent fountains in the great court before St. Peter's at Rome, which seem to form as essential a part of one architectural whole as the stately colonnade, or the massive dome itself. However strange, therefore, it may seem, some Daguerreotype [1] representations of the Falls have been executed with no small success. They not only record the form of the rocks and islands, but even the leading features of the cataract, and the shape of the clouds of spray. I often wished that Father Hennepin [2] could have taken one of these portraits, and bequeathed it to the geologists of our times. It would have afforded us no slight aid in our speculations respecting the comparative state of the ravine in the 19th and 17th centuries.

After one or two warm days, the weather became unusually cold for the month of June, with occasional frosts at night, and the humming-birds which we had seen before reaching Buffalo appeared no more during our stay here.

CHAPTER XX

JUNE 14, 1842.—From Queenstown we embarked in a fine steamer for Toronto, and had scarcely left the mouth of the river, and entered Lake Ontario, when we were surprised at seeing Toronto in the horizon, and the low wooded plain on which the town is built. By the effect of refraction, or "mirage," so common on this lake, the houses and trees were drawn up and lengthened vertically, so that I should have guessed them to be from 200 to 400 feet high, while the gently rising ground behind the town had the appearance of distant mountains. In the ordinary state of the atmosphere none of this land, much less the city, would be visible at this distance, even in the clearest weather.

Toronto contains already a population of 18,000 souls. The plain on which it stands has a gentle, and to the eye imperceptible, slope upwards from the lake, and is still covered for the most part, with a dense forest, which is beginning to give way before the axe of the new settler. I found Mr. Roy, the civil engineer, expecting me, and started with him the morning after my arrival.

In my ride with Mr. Roy through the forest we went about twenty miles due north of Toronto, besides making many detours. A more active scene of the progress of a new colony could scarcely be witnessed. We often came upon a party of surveyors, or pioneers, tracing out a new line of road with the trunks of tall trees felled on every side, over which we had to leap our horses. Then we made a circuit to get to windward of some large stumps which were on fire, or, if we could find no pathway, hurried our steeds through the smoke, half suffocated and oppressed with the heat of the burning timber and a sultry sun. Sometimes we emerged suddenly into a wide clearing, where not a single clump of trees had been spared by the impatient and improvident farmer. All were burnt, not even a shrub remaining for the cattle and sheep, which, for want of a better retreat were gasping under the imperfect shade of a wooden paling, called in America a Virginia, or snake fence.

The appearance of the country had been so entirely altered since Mr. Roy surveyed the ground two years before, and marked out the boundaries of the new settlements, that he lost his way while explaining to me the geology of "the ridges"; and after we had been on horseback for twelve hours we wandered about in a bright moonlight, unable to find the tavern where we hoped to pass the night. In the darker shade of the forest I saw many fire-flies; and my attention was kept alive, in spite of fatigue, by stories of men and horses swallowed up in some of the morasses which we crossed.

I shall always, in future, regard a corduroy road with
respect, as marking a great step in the march of civilization; for greatly were we rejoiced when we discovered in the moonlight the exact part of a bog, over
which a safe bridge of this kind had been laid down.
At length we reached a log-house, and thought our
troubles at an end. But the inmates, though eager to
serve us, could not comprehend a syllable of our language. I tried English, French, and German, all in
vain. Tired and disappointed, we walked to another
log-house, a mile farther on, leading our weary horses,
and then to others, but with no better success. Though
not among Indians, we were as foreigners in a strange
land. At last we stumbled, by good luck, upon our
inn, and the next day were told that the poor settlers
with whom we had fallen in the night before had all
come from the British Isles in the course of the five
preceding years. Some of them could speak Gaelic,
others Welsh, and others Irish; and the farmers were
most eloquent in descanting on their misfortune in
having no alternative but that of employing labourers
with whom they were unable to communicate, or remaining in want of hands while so many were out of
work, and in great distress. For the first time I became fully aware how much the success and progress
of a new colony depends on the state of schools in the
mother country.

CHAPTER XXI

JUNE 18.—An excellent steam-packet carried us along the northern coast of Lake Ontario, from Toronto to Kingston, from whence I made a geological excursion to Gannanoquoi. From Kingston we then descended the St. Lawrence to Montreal. The scenery of the Thousand Islands and of the rapids of the St. Lawrence owe much of their beauty to the clearness of the waters, which are almost as green, and their foam as white, as at the Falls of Niagara.

On approaching Montreal we seemed to be entering a French province. The language and costume of the peasants and of the old beggars, the priests with their breviaries, the large crosses on the public roads, with the symbols of the Crucifixion, the architecture of the houses, with their steep roofs, large casement windows, and, lastly, the great Catholic cathedral rising in state, with its two lofty towers, carried back our thoughts to Normandy and Brittany, where we spent the corresponding season of last year. The French spoken in those provinces of the mother country is often far less

156

correct, and less easy to follow, that that of the Canadians, whose manners are very prepossessing, much softer and more polite than those of their Anglo-Saxon Fellow-countrymen, however superior the latter may be in energy and capability of advancement.

Quebec, with its citadel and fortifications crowning the precipitous heights which overhang the St. Lawrence, and where the deep and broad river is enlivened with a variety of shipping, struck us as the most picturesque city we had seen since we landed in America. We were glad to meet with some old friends among the officers of the garrison, who accompanied us to the Falls of Montmorency, and other places in the neighborhood. Their task in maintaining discipline in their corps, in preventing the desertion of soldiers, and keeping the peace along the frontier, has been more irksome than in quelling the rebellion.

July 5th.—Returning to Montreal after our excursion to Quebec, we crossed the St. Lawrence on our way southward to La Prairie. On looking back over the river at Montreal, the whole city seemed in a blaze of light, owing to the fashion here of covering the houses with tin, which reflected the rays of the setting sun, so that every roof seemed a mirror. Behind the city rose its steep and shapely mountain, and in front were wooded islands, and the clear waters of the St. Lawrence sweeping along with a broad and rapid current. At the barracks in La Prairie, a regiment of hussars was exercising—a scene characteristic of the times. On the way to Lake Champlain we slept at St. John's, where I counted under the eaves of the stable of our inn more than forty nests of a species of swallow with a red breast. The head of a young bird was peeping out of each nest, and the old ones were flying about, feeding them. The landlord told me, that they had built there for twenty years, but missed the two years when the cholera raged, for at that time there was a scarcity of insects. Our host also mentioned, that in making an excavation lately near Prattsburg, about 1000 of these birds were found hybernating in the sand; a tale for the truth of which I do not vouch; but it agrees with some old accounts of the occasional hybernation of our swallows in similar situations.

We next crossed Lake Champlain to Burlington, in Vermont, in a steamboat, which, for neatness, elegance, and rapidity, excelled any we had yet beheld. The number of travellers has been sensibly thinned this year by the depressed state of commerce. The scenery of this lake is deservedly much admired. To the west we saw the principal range of mountains in the State of New York, Mount Marcy, the highest, attaining an elevation of upwards of 5400 feet. It is still (July 6th) capped with snow, but the season is unusually late.

July 9th.—From Burlington I crossed the Green Mountains of Vermont, passing by Montpelier, to Hanover. Here we paid a visit to Professor Hubbard, at Dartmouth College, and then returned through New Hampshire by Concord to Boston. Since we had left that city in May, we had travelled in little more than two months a distance of 2500 miles on railways,

in steamboats, and canoes, in public and private carriages, without any accident, and having always found
it possible so to plan our journey from day to day, as to
avoid all fatigue and night travelling. We had usually
slept in tolerable inns, and sometimes in excellent
hotels in small towns, and had scarcely ever been
interrupted by bad weather. I infer, from the dismay
occasionally expressed by Americans when we pursued
our journey, in spite of rain, that the climate of the
States must be always as we found it this year—won-
derfully more propitious to tourists than that of the
"old country," though it is said to be less favourable
to the health and complexion of Europeans.

I ventured on one or two occasions in Canada, when
I thought that the inns did not come up to the reasonable expectations of a traveller, to praise those of the
United States. I was immediately assured that if in
their country men preferred to dine at ordinaries, or
to board with their families at taverns, instead of cultivating domestic habits like the English, nothing
would be more easy than to have fine hotels in small
Canadian towns. This led me to inquire how many
families, out of more than fifty whom we had happened
to visit in our tour of eleven months in the United
States, resided in boarding-houses I found that there
was not one; and that all of them lived in houses of
their own. Some of these were in the northern and
middle, others in the southern and western states;
some in affluent, others in very moderate circumstances; they comprised many merchants as well as

lawyers, ministers of religion, political, literary, and scientific men.

Families who are travelling in the U. S., and strangers like ourselves, frequent hotels much more than in England, from the impossibility of hiring lodgings. In the inns, however, good private apartments may be obtained in all large towns, which, though dear for the United States, are cheap as contrasted with hotels in London. It is doubtless true that not only bachelors, but many young married couples, occasionally escape from the troubles of house-keeping in the United States, where servants are difficult to obtain, by retreating to boarding-houses; but the fact of our never having met with one instance among our own acquaintances inclines me to suspect the custom to be far less general than many foreigners suppose.

It is now the fourth time we had entered Boston, and we were delighted again to see our friends, some of whom kindly came from their country residences to welcome us. Others we visited in Nahant, where they had retreated from the great heat, to enjoy the sea-breezes. The fire-flies were rejoicing in the warm evenings. Ice was as usual in abundance; the iceman calling as regularly at every house in the morning as the milkman. Pine-apples from the West Indies were selling in the streets in wheelbarrows. I bought one of good size, and ripe, for a shilling, which would have cost twelve shillings or more in London. After a short stay, we set sail in the Caledonia steam-packet for Halifax.

CHAPTER XXIII

Halifax—High Tides in the Bay of Fundy—Progress and Resources of Nova Scotia—Promotion of Science—Nova Scotians "going home"—Return to England.

July 16, 1842.—When I went on board the Caledonia at Boston, I could hardly believe that it was as large as the Acadia, in which we had crossed the Atlantic from Liverpool, so familiar had I now become with the greater dimensions of the steamers which navigate the Hudson and other large American rivers.

The day after my arrival in Nova Scotia, a fellow-passenger in the coach from Halifax to Windsor, a native of the country, and who, from small beginnings, had acquired a large fortune, bore testimony to the rapid strides which the province had made, within his recollection, by deploring the universal increase of luxury. He spoke of the superior simplicity of manners in his younger days, when the wives and daughters of farmers were accustomed to ride to church, each on horseback behind their husbands and fathers, whereas now they were not content unless they could ride there in their own carriage.

In spite of the large extent of barren and siliceous soil in the south, and, what is a more serious evil, those seven or eight months of frost and snow which crowd the labours of the agriculturist into so brief a season, the resources of this province are extremely great. They have magnificent harbours, and fine navigable estuaries, large areas of the richest soil gained from the sea, vast supplies of coal and gypsum, and abundance of timber.

Not a few of the most intelligent and thriving inhabitants are descended from loyalists, who fled from the United States at the time of the declaration of independence.

I had arranged with Captain Bayfield, whom I had not seen for many years, that we should meet at Pictou, and the day after my arrival there, his surveying ship, the Gulnare, sailed into the harbour. I spent a day on board that vessel, and we then visited together the Albion Mines, from whence coal is conveyed by a railway to the estuary of the East River, and there shipped. Mr. Richard Brown, whose able co-operation in my geological inquiries I have before acknowledged, had come from Cape Breton to meet me, and with him and Mr. Dawson I examined the cliffs of the East River, accompanied by the superintendent of the Albion Mines, Mr. Poole, at whose house we were most kindly received. Here, during a week of intense heat, in the beginning of August (1842), I was frequently amused by watching the humming-birds, being able to approach unperceived, by the aid of a Venetian blind, to within a few inches of them, while they were on the wing. They remained for many seconds poised in the air, while sucking the flowers of several climbers trailed to the wall on the outside of the window, and in this position the head and body appeared motionless, brilliant with green and gold plumage, and the wings' invisible, owing to the rapidity of their motion. It is wonderful to reflect on the migrating instinct which leads these minute creatures from the distant Gulf of Florida to a country buried constantly under deep snow for seven or eight months in the year.

After leaving Pictou, I made an expedition with Mr. Dawson, and at Truro we were joined by Mr. Duncan, by whose advice we started at an early hour each morning in a boat, after the great tidal wave or bore had swept up the estuary, and were then carried ten, fifteen or twenty miles with great rapidity up the river, after which as the tide ebbed, we came down at our leisure; landing quietly wherever we pleased, at various points where the perpendicular cliffs offered sections on the right or left bank.

On one occasion, when I was seated on the trunk of a fallen tree, on a steep sloping beach about ten feet above the level of the river, I was warned by my companion that, before I had finished my sketch, the tide might float off me and the tree, and carry both down to the Basin of Minas. Being incredulous, I looked at my watch, and observed that the water remained nearly stationary for the first three minutes, and then, in the next ten, rose about three feet, after which it gained very steadily but more slowly, till I was obliged to decamp. A stranger, when he is looking for shells on the beach at low tide, after the hot sun has

nearly dried up the sandy mud, may well be surprised
if told that in six hours there will be a perpendicular
column of salt water sixty feet high over the spot on
which he stands.

The proprietor of one of the large gypsum quarries
showed me some wooden stakes, dug up a few days
before by one of his labourers from a considerable
depth in a peat bog. His men were persuaded that
they were artificially cut by a tool, and were the relics
of aboriginal Indians; but having been a trapper of
beavers in his younger days, he knew well that they
owed their shape to the teeth of these creatures. We
meet with the skulls and bones of beavers in the fens
of Cambridgeshire, and elsewhere in England. May
not some of the old tales of artificially cut wood occurring at great depths in peat and morasses, which
have puzzled many a learned antiquary, admit of the
like explanation?

I never travelled in any country where my scientific
pursuits seemed to be better understood, or were more
zealously forwarded, than in Nova Scotia, although I
went there almost without letters of introduction. At
Truro, having occasion to go over a great deal of
ground in different directions, on two successive days,
I had employed two pair of horses, one in the morning
and the other in the afternoon. The postmaster, an
entire stranger to me, declined to receive payment for
them, although I pressed him to do so, saying that
he heard I was exploring the country at my own expense, and he wished to contribute his share towards

scientific investigations undertaken for the public good.

We know, on the authority of the author of "Sam Slick," unless he has belied his countrymen, that some of the Blue Noses (so called from a kind of potato which thrives here) are not in the habit of setting a very high value, either on their own time or that of others. To this class, I presume, belonged the driver of a stage-coach, who conducted us from Pictou to Truro. Drawing in the reins of his four horses, he informed us that there were a great many wild raspberries by the road-side, quite ripe, and that he intended to get off and eat some of them, as there was time to spare, for he should still arrive in Truro by the appointed hour. It is needless to say that all turned out, as there was no alternative but to wait in the inside of a hot coach, or to pick fruit in the shade. Had the same adventure happened to a traveller in the United States, it might have furnished a good text to one inclined to descant on the inconvenient independence of manners which democratic institutions have a tendency to create. Doubtless, the political and social circumstances of all new colonies promote a degree of equality which influences the manners of the people. There is here no hereditary aristocracy—no proprietors who can let their lands to tenants—no dominant sect, with the privileges enjoyed by a church establishment. The influence of birth and family is scarcely felt, and the resemblance of the political and social state of things to that in the United States is striking.

It is no small object of ambition for a Nova Scotian to "go home," which means to "leave home, and see England." However much his curiosity may be gratified by the tour, his vanity, as I learn from several confessions made to me, is often put to a severe trial. It is mortifying to be asked in what part of the world Nova Scotia is situated—to be complimented on "speaking good English, although an American"—to be asked "what excuse can possibly be made for repudiation"—to be forced to explain to one fellow countryman after another "that Nova Scotia is not , one of the United States, but a British province." All this, too, after having prayed loyally every Sunday for Queen Victoria and the Prince of Wales—after having been so ready to go to war about the Canadian borderers, the New York sympathisers, the detention of Macleod and any other feud!

Nations know nothing of one another—most true—but unfortunately in this particular case the ignorance is all on one side, for almost every native of Nova Scotia knows and thinks a great deal about England. It may, however, console the Nova Scotian to reflect, that there are districts in the British Isles, far more populous than all his native peninsula, which the majority of the English people never heard of, and respecting which, if they were named, few could say whether they spoke Gaelic, Welsh, or Irish, or what form of religion the greater part of them professed.

August 18.—We left Halifax in the steamship Columbia, and in nine days and sixteen hours were at the pier at Liverpool. This was the ninetieth voyage of these Halifax steamers across the Atlantic, without any loss, and only one case of detention by putting back for repairs. As we flew along in the railway carriage between Liverpool and London, my eye, so long accustomed to the American landscape, was struck with the dressy and garden-like appearance of all the fields, the absence of weeds, and the neatness of the trim hedgerows. We passed only one unoccupied piece of ground, and it was covered with heath, then in full blossom, a plant which we had not seen from the time we crossed the Atlantic. Eight hours conveyed us from sea to sea, from the estuary of the Mersey to that stream which Pope has styled "the Father of the British Floods." Whatever new standard for measuring the comparative size of rivers I had acquired in my late wanderings, I certainly never beheld "the swelling waters and alternate tides" of Father Thames with greater admiration than after this long absence, or was ever more delighted to find myself once more in the midst of the flourishing settlement which has grown up upon his banks.

THE END.

NOTES

15, 1. Lyell wrote of his leaving Liverpool in 1845· Embarked with my wife at Liverpool, in the Britannia, one of the Cunard line of steamships, bound for Halifax and Boston On leaving the wharf, we had first been crammed, with a crowd of passengers and heaps of luggage, into a diminutive steamer, which looked like a toy by the side of the larger ship of 1200 tons, in which we were to cross the ocean I was reminded, however, by a friend, that this small craft was more than three times as large as one of the open caravels of Columbus, in his first voyage, which was only 15 tons burden, and without a deck. It is, indeed, marvelous to reflect on the daring of the early adventurers, for Frobisher, in 1576, made his way from the Thames to the shores of Labrador with two small barks of 20 and 25 tons each, not much surpassing in size the barge of a man-of-war; and Sir Humphrey Gilbert crossed to Newfoundland, in 1583, in a bark of 10 tons only, which was lost in a tempest on the return voyage. (*Second Visit*)

17, 1. Geological terms occur occasionally in this work. **Strata** are beds of earth and rock of one kind, formed by natural causes, each usually consisting of a series of layers.

18, 1. Sam Slick. A humorous book by Thomas C. Haliburton of Nova Scotia, in which the hero, Sam Slick, exaggerates the peculiarities of the Yankee character and dialect.

2. To us the most novel feature in the architectural aspect of the city, was the Bunker Hill Monument, which had been erected since 1842, the form of which, as it resembles an Egyptian obelisk, and possibly because I had seen the form imitated in some of our tall factory chimneys, gave me no pleasure. (Lyell, *Second Visit*)

19, 1. From Lyell's *Second Visit* The charge for the distance of fifty-four miles, from Boston to Portsmouth, was $1.50 each, or 6s. 4d. English, which was just half what we had paid three weeks before for first-class places on our journey from London to Liverpool (2l. 10s. for 210 miles) the speed being in both cases the same.

2. In the construction and management of railways, the Americans have in general displayed more prudence and economy than could have been expected, where a people of such sanguine temperament were entering on so novel a career of enterprise Annual dividends of seven or eight per cent. have been returned for a large part of the capital laid out on the New England railways, and on many others in the northern states. The cost of passing the original bills through the state parliaments has usually been very moderate, and never exorbitant; the lines have been carried as much as possible through districts where land was cheap, a single line only laid down where the traffic did not justify two, high gradients resorted to, rather than incur the expense of deep cuttings; tunnels entirely avoided; very little money spent in building station-houses; and except where the population was large, they have been content with the speed of fourteen or sixteen miles an hour. It has, moreover, been an invariable maxim "to go for numbers," by lowering the fares so as to bring them within the reach of all classes. Occasionally when the intercourse between two rich and populous cities, like New York and Boston, has excited the eager competition of rival companies, they have accelerated the speed far beyond the usual average; and we were carried from one metropolis to the other, a distance of 239 miles, at the rate of thirty miles an hour, in a commodious, lofty, and well-ventilated car, the charge being only three dollars, or thirteen shillings. (Lyell, *Second Visit*.)

20, 1. Observe the use of the word "avenues."

22, 1. Western steamers, notwithstanding their size, drew very little water, for they are constructed for rivers which rise and fall very rapidly. They cannot quite realize the boast of a western captain, "that he could sail wherever it was damp" but I was assured that some of them could float in two feet of water. (Lyell, *Second Visit*)

2. Basalt is usually of a dark color There are immense beds of it in some regions; and at the Giant's Causeway, on the north of Ireland, the columns are distinctly formed.

37, 1. Carboniferous. Coal-bearing.

39, 1. This incident illustrates one phase of the prevailing financial disturbance.

2. Among the most common singularities of expression are the following "I should admire to see him" for "I should like to see him;" "I want to know" and "Do tell," both exclamations of surprise, answering to our "Dear me" These last, however, are rarely heard in society above the middling class. Occasionally I was as much puzzled as if I was reading *Tam o'Shanter* as, for example, "out of kittel" means "out of order." (Lyell *Second Visit*)

40, 1. Maple sugar is made usually in the early spring.

55, 1. When I asked how it happened that in so populous and rich a city as Boston there was at present (October, 1845) no regular theatre, I was told, among other reasons, that if I went into the houses of persons of the middle and even humblest class, I should often find the father of a family, instead of seeking excitement in a shilling gallery, reading to his wife and four or five children, one of the best modern novels, which he had purchased for twenty-five cents, whereas, if they could all have left home, he could not for many times that sum have taken them to the play. They often buy, in two or three successive numbers of a penny newspaper, entire reprints of the tales of Dickens, Bulwer, or other popular writers. (Lyell, *Second Visit.*)

58, 1. Quoted from Virgil's *Æneid:* "What the Greeks and the merciless Achilles left."

68, 1. Quotation from Virgil's *Æneid:* "As the cypresses are wont to tower above the modest shrubs."

2. American histories give still another reason for the choice of a site on the Potomac for the national capitol.

98, 1. Mount Vernon is now maintained in excellent condition.

108, 1. Among other novelties since 1841, we observe with pleasure the new fountains in the midst of the city supplied from the Croton waterworks, finer than any which I remember to have seen in the centre of a city since I was last in Rome.

Among the new features of the city we see several fine churches, some built from their foundations, others finished since 1841. The two most conspicuous of the new edifices are Episcopalian, Trinity and Grace Church. The position of Trinity Church is admirably chosen, as it forms a prominent feature in Broadway, the principal street, and in another direction looks down Wall street, the great centre of city business. It is therefore seen from great distances in this atmosphere, so beautifully clear even at this season, when every stove is lighted, and when the thermometer has fallen twenty degrees below the freezing point.

Next to the new churches and fountains, the most striking change observable in the streets of New York since 1841, is the introduction of the electric telegraph, the posts of which, about 30 feet high and 100 yards apart, traverse Broadway, and are certainly not ornamental. . . . I learned that the length of line completed in 1846, amounted to above 1600 miles, and in 1848 there were more than 5000 miles of wire laid down. (Lyell, *Second Visit.*)

136, 1. Consult Dictionary for definition of "tropical"

138, 1. At Mr. Longworth's we saw a beautiful piece of sculpture, an ideal head called Ginevra, by Hiram Powers, who

had sent it from Rome as a present to his first patron. It ap
peared *to me worthy of the genius of the sculptor of "Eve" and*
the "Greek Slave." Thorwaldsen, when he saw Powers' "Eve,"
foretold that he would create an era in his art, and not a few of
the Italians now assign to him the first place in the "Natu
ralista" school, though assuredly there is much of the ideal also
in his conceptions of the beautiful. It augurs well for the futur
cultivation of the fine arts in the United States, that the Ameri
cans are as proud of their countryman's success as he himsel
could desire. (Lyell, *Second Visit.*)

142, 1. **Equus caballus.** Horse.

149, 1. Compare Tucker's estimates with the actual censu
returns. Population of the United States: 1850, 23,191,876
1860, 31,443,321; 1870, 38,558,371; 1880, 50,155,783; 1890
62,622,250; 1900, 76,303,387.

150, 1. **Carburetted hydrogen.** A compound of carbon and
hydrogen forming an illuminating gas.

152, 1. **Daguerreotype.** An early kind of photograph
named after its inventor.

2. **Father Hennepin.** A noted French missionary and ex
plorer of the Great Lakes and Upper Mississippi.

www.bookjungle.com *email: sales@bookjungle.com fax: 630-214-0564 mail: Book Jungle PO Box 2226 Champaign, IL 61825*

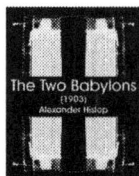

The Two Babylons
Alexander Hislop
You may be surprised to learn that many traditions of Roman Catholicism in fact don't come from Christ's teachings but from an ancient Babylonian "Mystery" religion that was centered on Nimrod, his wife Semiramis, and a child Tammuz. This book shows how this ancient religion transformed itself as it incorporated Christ into its teachings....

Religion/History   Pages:358

ISBN: *1-59462-010-5*   MSRP *$22.95*

QTY

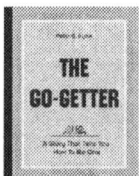

The Go-Getter
Kyne B. Peter
The Go Getter is the story of William Peck. He was a war veteran and amputee who will not be refused what he wants. Peck not only fights to find employment but continually proves himself more than competent at the many difficult test that are throw his way in the course of his early days with the Ricks Lumber Company...

Business/Self Help/Inspirational   Pages:68

ISBN: *1-59462-186-1*   MSRP *$8.95*

QTY

The Power Of Concentration
Theron Q. Dumont
It is of the utmost value to learn how to concentrate. To make the greatest success of anything you must be able to concentrate your entire thought upon the idea you are working on. The person that is able to concentrate utilizes all constructive thoughts and shuts out all destructive ones...

Self Help/Inspirational   Pages:196

ISBN: *1-59462-141-1*   MSRP *$14.95*

Self Mastery
Emile Coue
Emile Coue came up with novel way to improve the lives of people. He was a pharmacist by trade and often saw ailing people. This lead him to develop autosuggestion, a form of self-hypnosis. At the time his theories weren't popular but over the years evidence is mounting that he was indeed right all along...

New Age/Self Help   Pages:98

ISBN: *1-59462-189-6*   MSRP *$7.95*

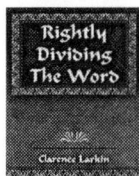

Rightly Dividing The Word
Clarence Larkin
The "Fundamental Doctrines" of the Christian Faith are clearly outlined in numerous books on Theology, but they are not available to the average reader and were mainly written for students. The Author has made it the work of his ministry to preach the "Fundamental Doctrines." To this end he has aimed to express them in the simplest and clearest manner..

Religion   Pages:352

ISBN: *1-59462-334-1*   MSRP *$23.45*

The Awful Disclosures Of
Maria Monk
"I cannot banish the scenes and characters of this book from my memory. To me it can never appear like an amusing fable, or lose its interest and importance. The story is one which is continually before me, and must return fresh to my mind with painful emotions as long as I live..."

Religion   Pages:232

ISBN: *1-59462-160-8*   MSRP *$17.95*

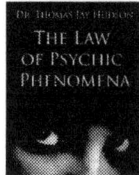

The Law of Psychic Phenomena
Thomson Jay Hudson
"I do not expect this book to stand upon its literary merits; for if it is unsound in principle, felicity of diction cannot save it, and if sound, homeliness of expression cannot destroy it. My primary object in offering it to the public is to assist in bringing Psychology within the domain of the exact sciences. That this has never been accomplished..."

New Age   Pages:420

ISBN: *1-59462-124-1*   MSRP *$29.95*

As a Man Thinketh
James Allen
"This little volume (the result of meditation and experience) is not intended as an exhaustive treatise on the much-written-upon subject of the power of thought. It is suggestive rather than explanatory, its object being to stimulate men and women to the discovery and perception of the truth that by virtue of the thoughts which they choose and encourage..."

Inspirational/Self Help   Pages:80

ISBN: *1-59462-231-0*   MSRP *$9.45*

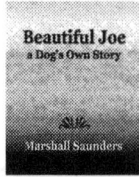

Beautiful Joe
Marshall Saunders
When Marshall visited the Moore family in 1892, she discovered Joe, a dog they had nursed back to health from his previous abusive home to live a happy life. So moved was she, that she wrote this classic masterpiece which won accolades and was recognized as a heartwarming symbol for humane animal treatment...

Fiction   Pages:256

ISBN: *1-59462-261-2*   MSRP *$18.45*

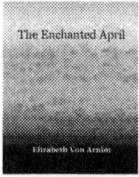

The Enchanted April
Elizabeth Von Arnim
It began in a woman's club in London on a February afternoon, an uncomfortable club, and a miserable afternoon when Mrs. Wilkins, who had come down from Hampstead to shop and had lunched at her club, took up The Times from the table in the smoking-room...

Fiction   Pages:368

ISBN: *1-59462-150-0*   MSRP *$23.45*

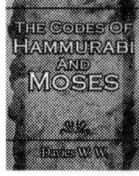

The Codes Of Hammurabi And
Moses - W. W. Davies
The discovery of the Hammurabi Code is one of the greatest achievements of archaeology, and is of paramount interest, not only to the student of the Bible, but also to all those interested in ancient history...

Religion   Pages:132

ISBN: *1-59462-338-4*   MSRP *$12.95*

Holland - The History Of Netherlands
Thomas Colley Grattan
Thomas Grattan was a prestigious writer from Dublin who served as British Consul to the US. Among his works is an authoritative look at the history of Holland. A colorful and interesting look at history....

History/Politics   Pages:408

ISBN: *1-59462-137-3*   MSRP *$26.95*

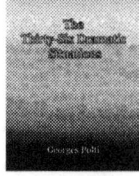

The Thirty-Six Dramatic Situations
Georges Polti
An incredibly useful guide for aspiring authors and playwrights. This volume categorizes every dramatic situation which could occur in a story and describes them in a list of 36 situations. A great aid to help inspire or formalize the creative writing process...

Self Help/Reference   Pages:204

ISBN: *1-59462-134-9*   MSRP *$15.95*

A Concise Dictionary of Middle English
A. L. Mayhew
Walter W. Skeat
The present work is intended to meet, in some measure, the requirements of those who wish to make some study of Middle-English, and who find a difficulty in obtaining such assistance as will enable them to find out the meanings and etymologies of the words most essential to their purpose...

Reference/History   Pages:332

ISBN: *1-59462-119-5*   MSRP *$29.95*

www.bookjungle.com email: sales@bookjungle.com fax: 630-214-0564 mail: Book Jungle PO Box 2226 Champaign, IL 61825

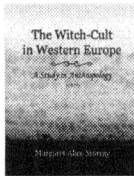

The Witch-Cult in Western Europe
Margaret Murray

The mass of existing material on this subject is so great that I have not attempted to make a survey of the whole of European "Witchcraft" but have confined myself to an intensive study of the cult in Great Britain. In order, however, to obtain a clearer understanding of the ritual and beliefs I have had recourse to French and Flemish sources...

QTY

Occult   Pages:308
ISBN: 1-59462-126-8   MSRP $22.45

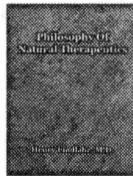

Philosophy Of Natural Therapeutics
Henry Lindlahr

We invite the earnest cooperation in this great work of all those who have awakened to the necessity for more rational living and for radical reform in healing methods...

QTY

Health/Philosophy/Self Help   Pages:552
ISBN: 1-59462-132-2   MSRP $34.95

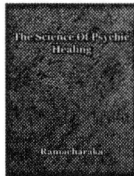

The Science Of Psychic Healing
Yogi Ramacharaka

This book is not a book of theories it deals with facts. Its author regards the best of theories as but working hypotheses to be used only until better ones present themselves. The "fact" is the principal thing the essential thing to uncover which the tool, theory, is used...

New Age/Health   Pages:180
ISBN: 1-59462-140-3   MSRP $13.95

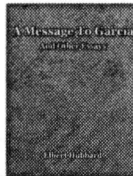

A Message to Garcia
Elbert Hubbard

This literary trifle, A Message to Garcia, was written one evening after supper, in a single hour. It was on the Twenty-second of February, Eighteen Hundred Ninety-nine, Washington's Birthday, and we were just going to press with the March Philistine...

New Age/Fiction   Pages:92
ISBN: 1-59462-144-6   MSRP $9.95

Bible Myths
Thomas Doane

In pursuing the study of the Bible Myths, facts pertaining thereto, in a condensed form, seemed to be greatly needed, and nowhere to be found. Widely scattered through hundreds of ancient and modern volumes, most of the contents of this book may indeed be found; but any previous attempt to trace exclusively the myths and legends...

Religion/History   Pages:644
ISBN: 1-59462-163-2   MSRP $38.95

The Book of Jasher
Alcuinus Flaccus Albinus

The Book of Jasher is an historical religious volume that many consider as a missing holy book from the Old Testament. Particularly studied by the Church of Later Day Saints and historians, it covers the history of the world from creation until the period of Judges in Israel. It's authenticity is bolstered due to a reference to the Book of Jasher in the Bible in Joshua 10:13

Religion/History   Pages:276
ISBN: 1-59462-197-7   MSRP $18.95

Tertium Organum
P. D. Ouspensky

A truly mind expanding writing that combines science with mysticism with unprecedented elegance. He presents the world we live in as a multi dimensional world and time as a motion through this world. But this isn't a cold and purely analytical explanation but a masterful presentation filled with similes and analogies...

New Age   Pages:356
ISBN: 1-59462-205-1   MSRP $23.95

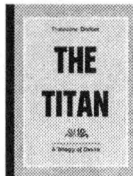

The Titan
Theodore Dreiser

"When Frank Algernon Cowperwood emerged from the Eastern District Penitentiary, in Philadelphia he realized that the old life he had lived in that city since boyhood was ended. His youth was gone, and with it had been lost the great business prospects of his earlier manhood. He must begin again..."

Fiction   Pages:564
ISBN: 1-59462-220-5   MSRP $33.95

Advance Course in Yogi Philosophy
Yogi Ramacharaka

"The twelve lessons forming this volume were originally issued in the shape of monthly lessons, known as "The Advanced Course in Yogi Philosophy and Oriental Occultism" during a period of twelve months beginning with October, 1904, and ending September, 1905."

Philosophy/Inspirational/Self Help   Pages:340
ISBN: 1-59462-229-9   MSRP $22.95

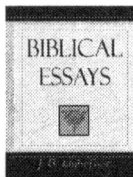

Biblical Essays
J. B. Lightfoot

About one-third of the present volume has already seen the light. The opening essay "On the Internal Evidence for the Authenticity and Genuineness of St John's Gospel" was published in the "Expositor" in the early months of 1890, and has been reprinted since...

Religion/History   Pages:480
ISBN: 1-59462-238-8   MSRP $30.95

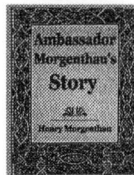

Ambassador Morgenthau's Story
Henry Morgenthau

"By this time the American people have probably become convinced that the Germans deliberately planned the conquest of the world. Yet they hesitate to convict on circumstantial evidence and for this reason all eye witnesses to this, the greatest crime in modern history, should volunteer their testimony..."

History   Pages:472
ISBN: 1-59462-244-2   MSRP $29.95

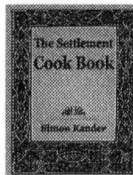

The Settlement Cook Book
Simon Kander

A legacy from the civil war, this book is a classic "American charity cookbook," which was used for fundraisers starting in Milwaukee. While it has transformed over the years, this printing provides great recipes from American history. Over two million copies have been sold. This volume contains a rich collection of recipes from noted chefs and hostesses of the turn of the century...

How-to   Pages:472
ISBN: 1-59462-256-6   MSRP $29.95

The Aquarian Gospel of Jesus the Christ
Levi Dowling

A retelling of Jesus' story which tells us what happened during the twenty year gap left by the Bible's New Testament. It tells of his travels to the far-east where he studied with the masters and fought against the rigid caste system. This book has enjoyed a resurgence in modern America and provides spiritual insight with charm. Its influences can be seen throughout the Age of Aquarius.

Religion   Pages:264
ISBN: 1-59462-321-X   MSRP $18.95

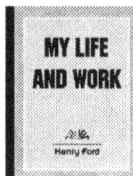

My Life and Work
Henry Ford

Henry Ford revolutionized the world with his implementation of mass production for the Model T automobile. Gain valuable business insight into his life and work with his own auto-biography... "We have only started on our development of our country we have not as yet, with all our talk of wonderful progress, done more than scratch the surface. The progress has been wonderful enough but..."

Biographies/History/Business   Pages:300
ISBN: 1-59462-198-5   MSRP $21.95

www.bookjungle.com email: sales@bookjungle.com fax: 630-214-0564 mail: Book Jungle PO Box 2226 Champaign, IL 61825

www.bookjungle.com *email: sales@bookjungle.com fax: 630-214-0564 mail: Book Jungle PO Box 2226 Champaign, IL 61825*

QTY

The Rosicrucian Cosmo-Conception Mystic Christianity *by Max Heindel*   ISBN: *1-59462-188-8* **$38.95**
The Rosicrucian Cosmo-conception is not dogmatic, neither does it appeal to any other authority than the reason of the student. It is: not controversial, but is: sent forth in the, hope that it may help to clear...   New Age/Religion Pages 646

Abandonment To Divine Providence *by Jean-Pierre de Caussade*   ISBN: *1-59462-228-0* **$25.95**
"The Rev. Jean Pierre de Caussade was one of the most remarkable spiritual writers of the Society of Jesus in France in the 18th Century. His death took place at Toulouse in 1751. His works have gone through many editions and have been republished...   Inspirational/Religion Pages 400

Mental Chemistry *by Charles Haanel*   ISBN: *1-59462-192-6* **$23.95**
Mental Chemistry allows the change of material conditions by combining and appropriately utilizing the power of the mind. Much like applied chemistry creates something new and unique out of careful combinations of chemicals the mastery of mental chemistry...   New Age Pages 354

The Letters of Robert Browning and Elizabeth Barret Barrett 1845-1846 vol II   ISBN: *1-59462-193-4* **$35.95**
by Robert Browning and Elizabeth Barrett   Biographies Pages 596

Gleanings In Genesis (volume I) *by Arthur W. Pink*   ISBN: *1-59462-130-6* **$25.95**
Appropriately has Genesis been termed "the seed plot of the Bible" for in it we have, in germ form, almost all of the great doctrines which are afterwards fully developed in the books of Scripture which follow...   Religion/Inspirational Pages 420

The Master Key *by L. W. de Laurence*   ISBN: *1-59462-001-6* **$30.95**
In no branch of human knowledge has there been a more lively increase of the spirit of research during the past few years than in the study of Psychology, Concentration and Mental Discipline. The requests for authentic lessons in Thought Control, Mental Discipline and...   New Age/Business Pages 422

The Lesser Key Of Solomon Goetia *by L. W. de Laurence*   ISBN: *1-59462-092-X* **$9.95**
This translation of the first book of the "Lernegton" which is now for the first time made accessible to students of Talismanic Magic was done, after careful collation and edition, from numerous Ancient Manuscripts in Hebrew, Latin, and French...   New Age/Occult Pages 92

Rubaiyat Of Omar Khayyam *by Edward Fitzgerald*   ISBN: *1-59462-332-5* **$13.95**
Edward Fitzgerald, whom the world has already learned, in spite of his own efforts to remain within the shadow of anonymity, to look upon as one of the rarest poets of the century, was born at Bredfield, in Suffolk, on the 31st of March, 1809. He was the third son of John Purcell...   Music Pages 172

Ancient Law *by Henry Maine*   ISBN: *1-59462-128-4* **$29.95**
The chief object of the following pages is to indicate some of the earliest ideas of mankind, as they are reflected in Ancient Law, and to point out the relation of those ideas to modern thought.   Religion/History Pages 452

Far-Away Stories *by William J. Locke*   ISBN: *1-59462-129-2* **$19.45**
"Good wine needs no bush, but a collection of mixed vintages does. And this book is just such a collection. Some of the stories I do not want to remain buried for ever in the museum files of dead magazine-numbers an author's not unpardonable vanity..."   Fiction Pages 272

Life of David Crockett *by David Crockett*   ISBN: *1-59462-250-7* **$27.45**
"Colonel David Crockett was one of the most remarkable men of the times in which he lived. Born in humble life, but gifted with a strong will, an indomitable courage, and unremitting perseverance...   Biographies/New Age Pages 424

Lip-Reading *by Edward Nitchie*   ISBN: *1-59462-206-X* **$25.95**
Edward B. Nitchie, founder of the New York School for the Hard of Hearing, now the Nitchie School of Lip-Reading, Inc, wrote "LIP-READING Principles and Practice". The development and perfecting of this meritorious work on lip-reading was an undertaking...   How-to Pages 400

A Handbook of Suggestive Therapeutics, Applied Hypnotism, Psychic Science   ISBN: *1-59462-214-0* **$24.95**
by Henry Munro   Health/New Age/Health Self-help Pages 376

A Doll's House: and Two Other Plays *by Henrik Ibsen*   ISBN: *1-59462-112-8* **$19.95**
Henrik Ibsen created this classic when in revolutionary 1848 Rome. Introducing some striking concepts in playwriting for the realist genre, this play has been studied the world over.   Fiction/Classics/Plays 308

The Light of Asia *by sir Edwin Arnold*   ISBN: *1-59462-204-3* **$13.95**
In this poetic masterpiece, Edwin Arnold describes the life and teachings of Buddha. The man who was to become known as Buddha to the world was born as Prince Gautama of India but he rejected the worldly riches and abandoned the reigns of power when...   Religion/History/Biographies Pages 170

The Complete Works of Guy de Maupassant *by Guy de Maupassant*   ISBN: *1-59462-157-8* **$16.95**
"For days and days, nights and nights, I had dreamed of that first kiss which was to consecrate our engagement, and I knew not on what spot I should put my lips..."   Fiction/Classics Pages 240

The Art of Cross-Examination *by Francis L. Wellman*   ISBN: *1-59462-309-0* **$26.95**
Written by a renowned trial lawyer, Wellman imparts his experience and uses case studies to explain how to use psychology to extract desired information through questioning.   How-to/Science/Reference Pages 408

Answered or Unanswered? *by Louisa Vaughan*   ISBN: *1-59462-248-5* **$10.95**
Miracles of Faith in China   Religion Pages 112

The Edinburgh Lectures on Mental Science (1909) *by Thomas*   ISBN: *1-59462-008-3* **$11.95**
This book contains the substance of a course of lectures recently given by the writer in the Queen Street Hall, Edinburgh. Its purpose is to indicate the Natural Principles governing the relation between Mental Action and Material Conditions...   New Age Psychology Pages 148

Ayesha *by H. Rider Haggard*   ISBN: *1-59462-301-5* **$24.95**
Verily and indeed it is the unexpected that happens! Probably if there was one person upon the earth from whom the Editor of this, and of a certain previous history, did not expect to hear again...   Classics Pages 380

Ayala's Angel *by Anthony Trollope*   ISBN: *1-59462-352-X* **$29.95**
The two girls were both pretty, but Lucy who was twenty-one who supposed to be simple and comparatively unattractive, whereas Ayala was credited, as her Bombwhat romantic name might show, with poetic charm and a taste for romance. Ayala when her father died was nineteen...   Fiction Pages 484

The American Commonwealth *by James Bryce*   ISBN: *1-59462-286-8* **$34.45**
An interpretation of American democratic political theory. It examines political mechanics and society from the perspective of Scotsman James Bryce   Politics Pages 572

Stories of the Pilgrims *by Margaret P. Pumphrey*   ISBN: *1-59462-116-0* **$17.95**
This book explores pilgrims religious oppression in England as well as their escape to Holland and eventual crossing to America on the Mayflower, and their early days in New England...   History Pages 268

BOOK JUNGLE

Bringing Classics to Life

www.bookjungle.com *email: sales@bookjungle.com fax: 630-214-0564 mail: Book Jungle PO Box 2226 Champaign, IL 61825*

QTY

The Fasting Cure *by Sinclair Upton*   ISBN: *1-59462-222-1*   **$13.95**
In the Cosmopolitan Magazine for May, 1910, and in the Contemporary Review (London) for April, 1910, I published an article dealing with my experiences in fasting. I have written a great many magazine articles, but never one which attracted so much attention... *New Age/Self Help/Health Pages 164*

Hebrew Astrology *by Sepharial*   ISBN: *1-59462-308-2*   **$13.45**
In these days of advanced thinking it is a matter of common observation that we have left many of the old landmarks behind and that we are now pressing forward to greater heights and to a wider horizon than that which represented the mind-content of our progenitors... *Astrology Pages 144*

Thought Vibration or The Law of Attraction in the Thought World   ISBN: *1-59462-127-6*   **$12.95**
by William Walker Atkinson   *Psychology/Religion Pages 144*

Optimism *by Helen Keller*   ISBN: *1-59462-108-X*   **$15.95**
Helen Keller was blind, deaf, and mute since 19 months old, yet famously learned how to overcome these handicaps, communicate with the world, and spread her lectures promoting optimism. An inspiring read for everyone... *Biographies/Inspirational Pages 84*

Sara Crewe *by Frances Burnett*   ISBN: *1-59462-360-0*   **$9.45**
In the first place, Miss Minchin lived in London. Her home was a large, dull, tall one, in a large, dull square, where all the houses were alike, and all the sparrows were alike, and where all the door-knockers made the same heavy sound... *Childrens/Classic Pages 88*

The Autobiography of Benjamin Franklin *by Benjamin Franklin*   ISBN: *1-59462-135-7*   **$24.95**
The Autobiography of Benjamin Franklin has probably been more extensively read than any other American historical work, and no other book of its kind has had such ups and downs of fortune. Franklin lived for many years in England, where he was agent... *Biographies/History Pages 332*

Name	
Email	
Telephone	
Address	
City, State ZIP	

☐ **Credit Card**   ☐ **Check / Money Order**

Credit Card Number	
Expiration Date	
Signature	

Please Mail to: Book Jungle
PO Box 2226
Champaign, IL 61825
or Fax to: 630-214-0564

ORDERING INFORMATION

web: *www.bookjungle.com*
email: *sales@bookjungle.com*
fax: *630-214-0564*
mail: *Book Jungle PO Box 2226 Champaign, IL 61825*
or PayPal *to sales@bookjungle.com*

Please contact us for bulk discounts

DIRECT-ORDER TERMS

**20% Discount if You Order
Two or More Books**
Free Domestic Shipping!
Accepted: Master Card, Visa,
Discover, American Express

www.ingramcontent.com/pod-product-compliance
Lightning Source LLC
Chambersburg PA
CBHW080530090426

42733CB00015B/2537